Birds of
Belize

Steve N. G. Howell and Dale Dyer

Princeton University Press

Princeton and Oxford

We dedicate this book to the genius Guy Tudor,
for his guidance of many years

Copyright © 2023 by Steve N. G. Howell and Dale Dyer

Princeton University Press is committed to the protection of copyright and the intellectual property our authors entrust to us. Copyright promotes the progress and integrity of knowledge. Thank you for supporting free speech and the global exchange of ideas by purchasing an authorized edition of this book. If you wish to reproduce or distribute any part of it in any form, please obtain permission.

Requests for permission to reproduce material from this work
should be sent to permissions@press.princeton.edu

Published by Princeton University Press
41 William Street, Princeton, New Jersey 08540
99 Banbury Road, Oxford OX2 6JX

press.princeton.edu

All Rights Reserved

Library of Congress Cataloging-in-Publication Data
Names: Howell, Steve N. G., author. | Dyer, Dale (Artist) illustrator.
Title: Birds of Belize / Steve N.G. Howell and Dale Dyer.
Description: First edition. | Princeton : Princeton University Press, [2023] | Series: Princeton field guides | Includes bibliographical references and index.
Identifiers: LCCN 2022029797 (print) | LCCN 2022029798 (ebook) | ISBN 9780691220727 (paperback) | ISBN 9780691220734 (ebook)
Subjects: LCSH: Birds—Belize—Identification. | BISAC: NATURE / Animals / Birds | TRAVEL / Special Interest / Ecotourism
Classification: LCC QL687.B45 H69 2023 (print) | LCC QL687.B45 (ebook) | DDC 598.097282—dc23/eng/20220712
LC record available at https://lccn.loc.gov/2022029797
LC ebook record available at https://lccn.loc.gov/2022029798

British Library Cataloging-in-Publication Data is available

Cover Credit: Keel-billed Toucans, by Dale Dyer

This book has been composed in Adobe Garamond Pro

Printed on acid-free paper. ∞

Typeset and designed by D & N Publishing, Wiltshire, UK

Printed in Italy

10 9 8 7 6 5 4 3 2 1

GAMEBIRDS AND ALLIES
pp. 84–88

Quail
p. 84

Cracids, Turkey
pp. 84–86

Tinamous
p. 88

RAPTORS AND OWLS
pp. 90–124

Vultures
p. 90

Hawks, Kites, Eagles
pp. 92–114

Falcons
pp. 92, 112,
116–118

Owls, Barn Owl
pp. 120–124

LARGER LANDBIRDS
pp. 124–160

Cuckoos, Anis
pp. 142–144

Potoos, Nightjars
pp. 124–128

Pigeons, Doves
pp. 130–136

Parrots
pp. 138–142

Jacamars
p. 152

Kingfishers
p. 148

Motmots
p. 150

Puffbirds
p. 150

Trogons
p. 146

Toucans
p. 152

Woodpeckers
pp. 154–158

Jays
p. 160

AERIAL LANDBIRDS
pp. 162–178

Hummingbirds
pp. 162–170

Swifts
pp. 172–174

Swallows
pp. 174–178

SONGBIRDS
pp. 180–286

Antbirds
pp. 186–188

Ovenbirds
pp. 180–184

Antthrush
p. 188

Cotingas
p. 194

Manakins
p. 192

Tityras, Becards, Mourners
pp. 190–194

Flycatchers
pp. 196–214

SONGBIRDS
pp. 180–286

Mimids
p. 216

Thrushes
pp. 216–220

Waxwing
p. 220

Wrens
pp. 220–224

Gnatwren, Gnatcatchers
p. 224

Vireos
pp. 226–230

Warblers
pp. 232–246

Chat
p. 248

Sparrows
pp. 248–252

Munias
p. 254

Finches
pp. 254–256

Grosbeaks, Buntings
pp. 258–266

Tanagers
pp. 268–276

Blackbirds, Orioles
pp. 276–286

CONTENTS

PREFACE AND ACKNOWLEDGMENTS

Belize, formerly British Honduras, is a small, English-speaking country fronting the Caribbean Sea and wedged into the Middle American isthmus south of Mexico's Yucatan Peninsula and east of Guatemala (see map on p. 8). Some 600 species of birds (in about 72 families) have been recorded in Belize, with a land area of about 23,000 km², smaller than all but four U.S. states.

For many years, birders in Belize had to make do with works covering much larger geographic areas, such as the hefty guide to Mexico and northern Central America by Steve Howell and Sophie Webb (1995), which by necessity treated Belize as a minor part of a bigger entity and could not do the country justice at a local level. In recent years, however, resident and visiting birders have been well served by the field guide authored by long-term Belize resident Lee Jones (2003), a guide built upon years of exploring the country, combined with a summary of the literature, notably the baseline work on avian distribution by Stephen M. Russell (1964). Belizean species are also treated in the more recent *Birds of Central America* (Vallely & Dyer 2018), the plates from which form the basis for the present work. We are fortunate to have had such a wealth of prior literature to build upon, and we thank all those involved in earlier works. In terms of labor division, Howell had primary responsibility for the text, and Dyer painted all the plates.

Howell first visited Belize in 1984, as part of a five-year odyssey traveling and birding throughout Mexico and Central America; he has visited many times since and led numerous tours to most parts of the country and to adjacent regions. As part of the mammoth task of illustrating the birds of Central America, Dyer first visited Belize in 1996 and then Belize and other Central American countries annually during 2007–2016. Between us and our many months of fieldwork in Belize and neighboring countries we have gained field experience with all but one of the species included in this guide (damn that elusive Harpy Eagle!).

Fieldwork is only part of the equation, however, and a book such as this is the product of many people. In particular, for their long-term support of our work we thank the personnel at WINGS, especially Will Russell and Matt Brooks; all at the Palomarin Field Station of Point Blue (formerly PRBO), especially Diana Humple, Mark Dettling, Megan Elrod, and Renée Cormier; and the personnel of the Ornithology Department at the American Museum of Natural History, New York, especially Paul Sweet, Bentley Bird, Peter Cappainolo, and curators Joel Cracraft, Brian Tilston Smith, and George Barrowclough.

Our taxonomic research was aided greatly by the herculean effort of the *Handbook of the Birds of the World* team in producing their illustrated checklists, including numerous vocal analyses by Peter Boesman; and the work of the North American Checklist Committee, South American Checklist Committee, and International Ornithologists' Union in maintaining their checklists and providing references. The sound archives of the Macaulay Library at the Cornell Lab of Ornithology (www.macaulaylibrary.org) and Xeno-canto (www.xeno-canto.org) are fantastic resources that also helped immeasurably with our work, and we thank the many thoughtful people who have uploaded recordings to these platforms.

For help in questions concerning status and distribution of Belizean birds we are indebted to Lee Jones for his frequent correspondence and attention to detail; he also kindly contributed the habitat photos in the introduction. For help in the field, we especially thank Bryan Bland, Stu Tingley, Andrew Vallely, and Sophie Webb. Dennis Jongsomjit of Point Blue helped greatly by creating the base maps we used for the introduction and for the species range maps. Others who helped in various ways include Michael O'Brien, J. Van Remsen, and Michael L. P. Retter. The book benefited from review by Lee Jones and Michael O'Brien, and any remaining errors are our responsibility. Last, but far from least, we thank the remarkable Robert Kirk and team at Princeton University Press, along with David Price-Goodfellow and D & N Publishing, for their care in bringing *Birds of Belize* to fruition.

HOW TO USE THIS BOOK

Area and Species Covered

In this guide to field identification, we cover bird species found regularly in Belize, including the numerous small islands of the barrier reef, plus marine waters that can be reached within a day trip (out to about 50km, or 30 miles, from shore). We don't include very rare and vagrant species unlikely to be seen in Belize (for completeness, these are listed in Appendix A). In a few cases, though, such as when a rare species is similar to a regularly occurring one, we have included rarer species on the plates to encourage clarification of status; in some cases, plate design considerations influenced the inclusion or omission of rarer species. Our baseline taxonomy is that of IOC (Gill et al. 2021), except for oceanic birds, where we follow Howell & Zufelt (2019). Our cut-off date for species coverage and inclusion is December 2021.

Format

The inside front cover and first three pages offer a pictorial contents to help you get to the right group of birds. Putting a bird into the correct group of species is an important first step in any identification process—if you are trying to identify a duck by looking among hawks, well, as they say, 'you can't get there from here.' Birds are arranged in a user-friendly field guide sequence (basically waterbirds, then landbirds) following Howell et al. (2012) rather than in a phylogenetic order, which is often unsuited to field identification (Yoon 2009).

Plates

We have tried to group species on plates by similarity in appearance integrated with shared geographic distribution and habitat—thus, similar species likely to be found together are usually on the same plate or on adjacent plates. Species on a plate are shown at the same scale, except in a few cases where a line divides the plate into different scales; this should be obvious by consulting the lengths given in the species accounts.

When different ages and sexes are shown, images for a species are arranged with juveniles and immatures on the left, adults on the right (usually females, then males). No label for age or sex indicates simply an 'adult' bird in which sexes look similar (many 1st-year songbirds are not distinguishable from adults after molting out of their briefly held juvenile plumage). When shown, juveniles and immatures are labeled as such, as are males and females. Well-defined seasonal plumages (mainly for northern migrants) are often followed by month spans to indicate when these plumages are typically seen.

Family and Genus Accounts

Each family, plus some genera and other species groups, start with a short account summarizing features common to the family or group. (The number of regularly occurring species in each family is noted in parentheses; a plus sign (+) indicates additional rarer species in that family have been recorded in Belize, noted in Appendix A).

Species Accounts

These start with English and scientific names, plus length (and sometimes wingspan) in cm; length is bill tip to tail tip, measured from museum specimens laid on their back with no undue stretching applied. An asterisk (*) preceding the species name refers you to taxonomic comments in Appendix B.

The use of parentheses in an English or scientific name indicates an alternative name, such as Bank Swallow (Sand Martin), where the name Sand Martin is used for this species in the Old World, or *Tangara (Thraupis) abbas* for Yellow-winged Tanager, where some authors place this species in the genus *Tangara*, others in the genus *Thraupis*. Because names are changing so frequently, we have used parentheses only for what we consider the most widely used alternate names. For example, Smoky-brown Woodpecker has been transferred among four (!) different genera in recent years, and different authorities still place it in at least two genera.

To convey the frequently ambiguous nature of taxonomy, we employ brackets in both the English and scientific names. These brackets indicate relatedness, or traditionally perceived relatedness, for taxa that

Political geography of Belize, showing political districts (indicated with all capital letters), towns, and other places mentioned in the text.

fall either into the gray zone of uncertainty or into the realm of proposed new splits. If a taxon has a widely used English name, or if its original name is evident from the new English name, we use brackets only in the scientific name, as in Northern Collared Trogon *Trogon [collaris] puella*, or Northern Social Flycatcher *Myiozetetes [similis] texensis*; the name in brackets is simply the chronologically first-named taxon, which may be called the parent species. If the taxon has a potentially unfamiliar English name we include the former (parent) English name in brackets, as with Yucatan [Caribbean] Dove *Leptotila [jamaicensis] gaumeri* or Black [Variable] Seedeater *Sporophila corvina*; we do this also for some relatively recent splits that are widely agreed upon but may be unfamiliar to some users, such as Gartered [Violaceous] Trogon *Trogon [violaceus] caligatus* and Morelet's [White-collared] Seedeater *Sporophila [torqueola] morelleti*. Although sometimes a little cumbersome, we believe that the information provided by this method is preferable to simply dumping a load of new and potentially confusing names on the reader with no explanation.

Lengths simply give an idea of relative size; note that every species varies slightly to strikingly in length (most books provide simply an 'average' length, which can be quite misleading). Moreover, because area is a square of length, a bird 10% 'larger' (= longer) than another can appear appreciably bigger in the field, e.g., a bird 10cm long can appear 20% 'bigger' than a bird with similar proportions that is 9cm long (10 × 10 = 100cm, vs. 9 × 9 = 81cm).]

The species account text then covers points relevant to identification (hereafter, ID), which can include aspects of habitat, seasonal occurrence, abundance, and behavior, as well as the more conventional ID criteria of plumage, structure, and voice. Habitat descriptions are usually broad-brush, and behavioral notes are generally limited to features that can assist with ID.

Sounds can be very helpful for ID. Although calls and songs are notoriously difficult to describe, voice descriptions may help in several ways: (1) to ID a bird from something seen poorly but heard well; (2) to prompt the memory, including comparison with similar sounds; (3) to tell you what a bird *does not* sound like, often helpful when seeking a species—don't follow-up on high lisping twitters if the bird you are seeking makes low-pitched rattles.

We use accents to indicate emphasis or a point of rising emphasis, as in *peé-eer*; and CAPITAL letters to indicate loud or strongly emphasized phrases, as in *PEE-eer*. We describe what we consider the most typical and frequently heard vocalizations; almost all species make myriad other sounds, especially in the breeding season. Sound descriptions are based on the personal experience of Howell, including reference to the online sound libraries of the Macaulay Library (www.macaulaylibrary.org) and Xeno-canto (www. xeno-canto.org). This section is omitted if a species is typically silent in Belize.

Although many sounds can be found online and on apps, beware of potential geographic as well as individual variation, not to mention the possibility of misidentified sound files (especially in the Macaulay Library, whose utility is being diluted by burgeoning poor-quality recordings added by well-meaning eBird users). Moreover, vocalizations on commercial CDs in particular can be of birds agitated in response to playback vs. birds giving vocalizations more typically heard in the field.

Status: Statements made here should be used in conjunction with the maps. Species are assumed to be resident unless stated otherwise. Migrants include *winter migrants* (mainly visitors from North America), *transient migrants* (passing through to/from North and South America), *summer migrants* (visiting Belize in summer to breed), and simply *nonbreeding (nonbr.) migrants* (such as wide-ranging waterbirds or shorebirds that may remain in Belize during their first summer rather than returning north to breed).

Date and elevational ranges are by necessity broad-brush guidelines. Thus, a bird described as occurring from mid-March to September would not be exceptional if encountered in early March or early October, whereas an occurrence in January would be highly unusual and should be documented carefully. Given that Belize is largely a low-lying country, elevation is not as relevant as in mountainous countries such as Mexico or Costa Rica; elevation is noted only when a species is typically found at or above certain elevations in the Maya Mountains. World range is summarized (in parentheses) at the end of the Status section.

Abundance terms are relative and averaged out over several years; migrants, especially transients, may be common one week, rare the next, depending on local weather conditions. Abundance also often varies regionally or by habitat and elevation; and simply by virtue of lifestyle, common hawks are numerically less numerous than common tanagers.

Common: encountered (seen or heard) on most or all days *in range and habitat/season*, often in relatively large numbers. For example, Black Vulture, Great Kiskadee, Blue-gray Tanager.

Fairly common: encountered (seen or heard) on most or all days *in range and habitat/season*, in smaller numbers than common species. For example, Double-toothed Kite, Mesoamerican [Greenish] Elaenia, Crimson-collared Tanager.

Uncommon: not usually encountered (seen or heard) on most days *in range and habitat/season*, and typically in small numbers. For example, Ornate Hawk-Eagle, Red Crossbill.

Scarce: species that appear rare, but perhaps due more to their behavior or to observer coverage than to actual rarity, such as Northern Black Rail, Bicolored Hawk; or to population reduction through habitat change and hunting, as for Great Curassow in much of its range.

Rare: species that occurs in low density, usually missed far more often than seen, even in range and habitat/season; perhaps only encountered a few times a year. For example, Solitary Eagle, Blue-headed Vireo.

Very rare: not usually encountered every year, and should be documented carefully. For example, Crested Eagle, Long-billed Starthroat.

Irregular: Abundance varies appreciably between years, not necessarily annual in occurrence. For example, Green-winged Teal, Cedar Waxwing.

Range Maps

Range maps are by necessity simplified and often extrapolated, but they should give a general idea of distribution and seasonal occurrence. The maps are based on peer-reviewed literature, specimens, our own field experience, and eBird (but see below). Basically, the maps show where a species may be expected to occur *within habitat and season* and should be used in conjunction with the Status sections of the species accounts. A quick visual check of, say, p. 114, shows immediately that Swallow-tailed Kite is a seasonal

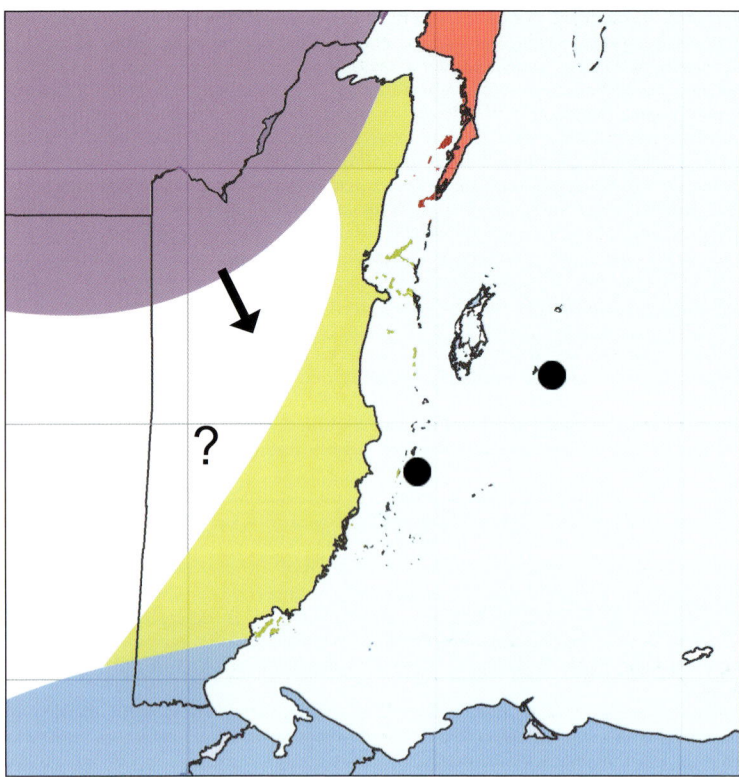

Key to species range maps.

Resident breeding species; breeding sites may be rather local for some species, especially colonial waterbirds

Seasonal breeding migrant, usually in summer; breeding sites may be rather local for some species, especially colonial waterbirds

Seasonal nonbreeding migrant, usually in winter but may be year-round for some species, especially waterbirds

Transient migrant

Breeding colony or isolated population

Occurrence possible or reported, status in need of elucidation or confirmation

Direction of range expansion

breeding migrant in the south and northwest, a transient elsewhere; White-tailed Kite is a widespread resident absent from heavy forest and higher elevations; Plumbeous Kite is a widespread seasonal breeding migrant; and Mississippi Kite is a transient migrant in the far south.

Because seasonal occurrence can be complex at a local level, and is poorly known in many cases, we have sometimes opted to map only the primary seasonal status, with notes in the text about local differences; for example, the text notes that post-breeding summer flocks of Swallow-tailed Kite may occur outside the mapped breeding range; and Mississippi Kite is mapped as a transient migrant only in the far south, but the text mentions it is rare and sporadic to the north of the mapped range.

Most resident tropical species do not wander much, but migrants might on occasion be found well outside the mapped range. Distribution is also dynamic, as for species whose ranges are expanding or contracting with habitat modification and climate change. Maps are not included for very rare migrants, and also do not show exceptional or vagrant records, but these may be summarized in the text. Thus, the lack of a range map usually indicates a species unlikely to be encountered in Belize.

A note on eBird (see www.ebird.org). Based on the infinite number of monkeys theorem, this popular resource has great potential and was extremely helpful in creating the range maps. However, for numerous taxa (from Mexico south to Costa Rica) we often found at least 10–20% of *documented* (by photo or sound recording) eBird reports to be misidentified. This is not a commentary on the gallant eBird reviewers, who have volunteered for this thankless, Sisyphean task, but simply an inevitable consequence of the sheer volume of unfiltered input far exceeding the foreseeable capacity for proofing. When our maps or range statements do not agree with eBird it is usually because we discount unverified or inaccurate reports rather than having overlooked them. Undoubtedly this will lead to errors of omission for some species, and we encourage users to document records outside the mapped ranges.

Abbreviations and Some Terms Explained

We have limited the use of abbreviations and have tried to use widely understood terms for geography, habitat, bird anatomy, and such. Nonetheless, a few terms may be unfamiliar to some users and are explained below.

We use cen. for central, and n., s., ne. for north, south, northeast, etc. (capitalized for major regions, as in N Atlantic or S America, etc.); I. for island, and Is. for islands. For months we use 3 letters: Jun = June, Sep = September, etc. For plumages we use nonbr. for nonbreeding, juv. for juvenile (a bird in its first non-downy = juvenile plumage), and imm. for immature (any non-adult plumage, including juvenile). We use WS for wingspan. Finally, in some places we use ID for identification and the Latin abbreviation cf. for compare with.

Bamboo Woody perennial grass that grows in dense mats or stands; flowers and seeds at long and unpredictable intervals.

Caye (often pronounced 'key') Low-lying coral limestone islet, typically with low vegetation dominated by mangroves.

Cere A bare leathery patch of skin into which the nostrils open; seen mainly on hawks and falcons, often brightly colored.

Cloud Forest Mature, humid evergreen forest growing mainly on ridgetops where clouds form and provide moisture throughout the year; often dense, loaded with epiphytes; can be stunted and mossy on windiest ridgetops, when sometimes termed Elfin Forest.

Gallery Forest Corridors of humid forest growing along watercourses, usually in otherwise dry or more open habitats, as may be seen along many rivers flowing through pine savanna.

Heliconia Genus of herbaceous flowering plants with long broad leaves (suggestive of banana leaves) growing in successional and swampy habitats, the ornate and colorful flowers attractive to hermits and other hummingbirds; also known as wild plantain or *platanillo*.

Holarctic Temperate regions of North America and Eurasia.

Lore(s) The area between a bird's eye and the base of its bill; dark lores or pale lores can greatly influence facial expression.

Humid Forest Mature, semi-deciduous and evergreen broadleaf forest, includes rainforest.

Mandible The lower half of the bill; sometimes called lower mandible.

Bird Topography

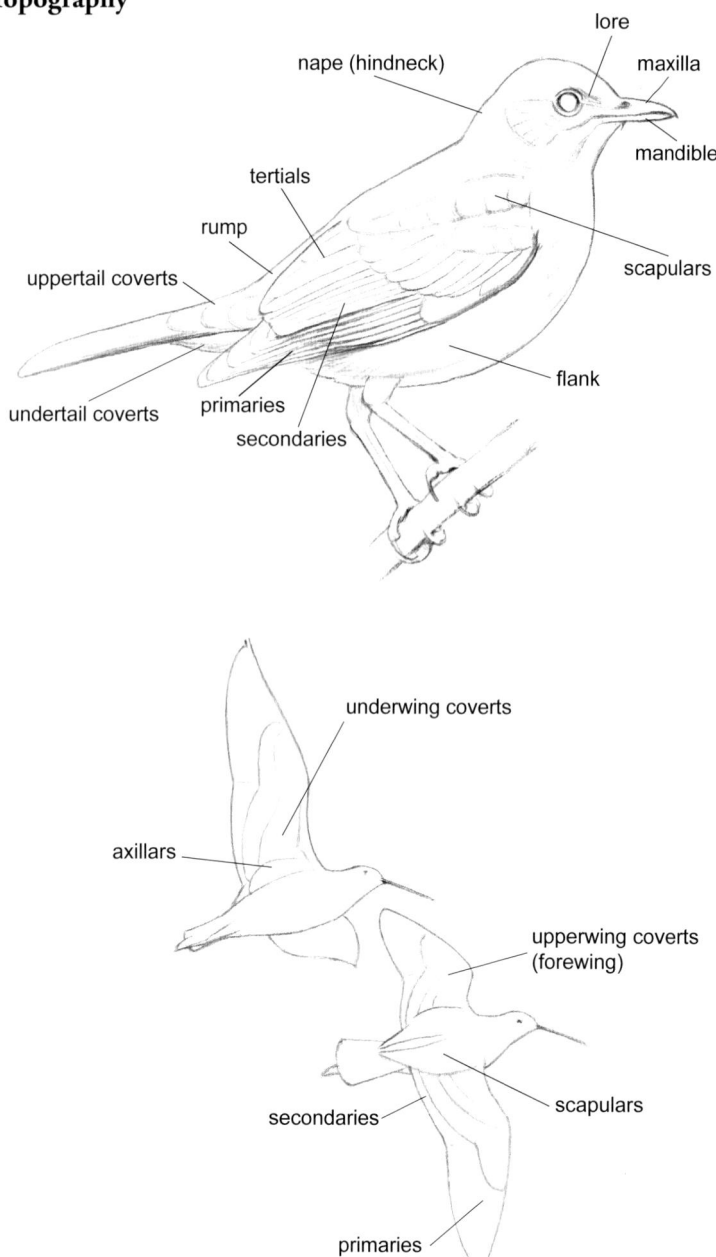

Most terms we use should be self-explanatory, especially in reference to the plates.
Some of the more important and potentially unfamiliar terms are illustrated above.

Mangroves Specialized plant community permanently or periodically inundated with salt or brackish water; can be low and scrubby or tall and forest-like, some species with stilt roots.

Maxilla The upper half of the bill; sometimes called upper mandible.

Neotropical Of New World tropical regions.

Pantropical Of tropical regions worldwide.

Rainforest Mature, tropical broadleaf lowland forest supported by rainfall throughout the year, without a marked dry season; although all trees lose their leaves at some time or other, forest retains a lush evergreen aspect year-round; trees are tall, often with buttress roots, and understory is often fairly open. Rainforest is typical of southern Belize, and grades into semi-deciduous forest across much of northern Belize.

Savanna Natural grassland with scattered trees and bushes, growing mainly on poorer soils; in Belize, pine savanna, with open pine woods, is a dominant feature of many lowland areas nearer the coast.

Second Growth Successional-stage shrubby and wooded habitats typical of disturbed areas, roadsides; trees generally smaller and canopy lower than in undisturbed forest.

Semi-deciduous Forest Broadleaf forest that appears evergreen but has a distinct if short dry season when many trees lose most or all of their leaves. Typical of much of the Yucatan Peninsula, including northern Belize, where it grades into rainforest.

Speculum Iridescent panel on the upper surface of the secondaries in various duck species.

BIOGEOGRAPHY

Birds and their habitats are not distributed randomly, and a basic understanding of how geography and climate shape the distribution of habitat types—and consequently bird distribution—can help you to both find and identify species. While there is still much to learn about the finer points of distribution and seasonal occurrence of birds in Belize, here we offer an overview of the geography, climate, and habitat to give a sense of bird distribution in this small but biologically rich country.

Geography

Nestled between Mexico and Guatemala at the eastern base of the Yucatan Peninsula and fronting the Caribbean Sea, Belize was known formerly as British Honduras (up until 1973) and was a British colony until independence in 1981. Centered at about 17°N and 89°W, at the northern edge of Central America, the country is oriented roughly from north to south, with its mainland dimensions only about 300 × 100km (180 × 65 miles) and a land area only slightly larger than New Jersey or Wales. Most of Belize is low-lying, below 300m elevation, with its two highest peaks only just over 1100m. But don't let these statistics fool you. Despite its small area, many areas of Belize comprise rugged and wild country; as recently as 1990 most main highways were unpaved, and access to higher elevations still represents a serious expedition.

The northern lowlands are limestone-based, forming the southern edge of the Yucatan Peninsula, a broad, low-lying shelf (mostly within Mexico) that projects north into the Gulf of Mexico. The only highland area, the Maya Mountains, is formed of volcanic and metamorphic rock and cuts across the southern half of Belize to somewhat split the country into northern and southern sections, each with a slightly different avifauna. Numerous rivers drain the landscape, and there are also several large freshwater lagoons in the northern half of the country. The coast of Belize is fringed by a long barrier reef dotted with numerous low-lying islets known as cayes.

Belize has the lowest human population of any Middle American country, with just over 400,000 people in 2019. This fact helps account for the large areas of remaining forest, and more than a third of the land territory has been given some form of protection, at least on paper. That said, deforestation is increasing rapidly as the human population burgeons—Belize has one of the highest human population growth rates in the Americas.

Climate and Habitat

Belize has a generally hot and humid tropical climate and lies in the latitude of the Northeast Trade Winds, which blow warm moist air onshore from the Caribbean Sea through most of the year. As the sun shifts north in summer, increased heating of the land causes this onshore-flowing air to rise and cool, dropping its moisture to produce a rainy season that in Belize overall spans June through November. The wet season often starts fairly abruptly with heavy showers in late May and tails off gradually into November. Hurricanes and tropical storms are not infrequent and supplement the regular wet season rain. Mean annual rainfall varies from about 1.2 meters in the north, along the Mexico border, to 4.5 meters in the south, around Punta Gorda.

The dry season spans roughly January to May in northern Belize, February to April in the south, although it is not strongly marked thanks to cold fronts from North America (known as 'northers') and frequent ground fog. From October into April (mainly November to early March), cold fronts moving south from the interior of North America bring cooler air and periods of rain, rarely up to a week, while frequent early morning fogs also provide considerable moisture and relief from potentially long dry periods. Hence, most forest in Belize retains an evergreen aspect, unlike the seasonally leafless, gray, and dead-looking forests of tropical regions with a pronounced dry season, such as much of the Pacific slope of Middle America.

As well as rainfall, soil type plays an important role in vegetation type. The shallow limestone soils in northern Belize, combined with lower rainfall, help explain the predominantly semi-deciduous forest there compared with the taller evergreen forest found south of the Maya Mountains. Many classic lowland Neotropical bird groups are well represented among birds typical of both northern and southern humid lowland forests, including tinamous, parrots, toucans, motmots, puffbirds, jacamars, woodcreepers, antbirds, and manakins.

Humid evergreen forest edge at Black Rock Lodge, Cayo. 24 April 2018. © *H. Lee Jones.*

Belize's northern forests have a distinctly Yucatan flavor, with species such as Yucatan Parrot, Yucatan Poorwill, Yucatan Woodpecker, Yucatan Flycatcher, Yucatan Jay, Rose-throated Tanager, and Northern Cardinal all at the edge of their ranges. Conversely, species reaching the northern edge of their range in southern Belize include Band-tailed Barbthroat, Mistletoe Tyrannulet, Bare-crowned Antbird, and Black-crowned Antshrike. The forests of southern Belize are still relatively poorly studied, and other resident species may yet be found there, such as Olivaceous Piculet *Picumnus olivaceus*, Scaled Antpitta *Grallaria guatimalensis*, and even perhaps Gray-headed Piprites *Piprites griseiceps*, all of which are generally inconspicuous and often occur at low density.

The other main lowland habitat in Belize is pine savanna, growing on poor alluvial soils washed down over time from the Maya Mountains. Much of eastern Belize is covered by pine savanna, dissected by rivers lined with taller trees and bamboo thickets. Species typical of this habitat include Yucatan Bobwhite, Aplomado Falcon, Yellow-headed Parrot, Ladder-backed Woodpecker, and Vermilion Flycatcher, with the bamboo specialist Blue Seedeater occurring locally along rivers.

Open pine forest also dominates a large section of the Maya Mountains, known as the Mountain Pine Ridge. Several bird species there are shared with the lowland pine savannas, such as Acorn Woodpecker, Grace's Warbler, Hepatic Tanager, and Chipping Sparrow, whereas others are found only in the highlands, including Greater Pewee, Black-headed Siskin, and Red Crossbill.

Rising warm air that cools along the peaks of the Maya Mountains year-round creates cloud and mist, which support local pockets of cloud forest. This higher elevation, humid evergreen forest has a distinctive avifauna that is difficult to access in Belize. Species mostly limited to this habitat include Brown Violetear, Northern Spotted Woodcreeper, Scaly-throated Foliage-gleaner, Tawny-throated Leaftosser, Middle American Bush-tanager, and Flame-colored Tanager. In some years, presumably reflecting food supply, a few fruit-eating species breeding in the Maya Mountains may disperse to the adjacent lowlands, such as Slate-colored Solitaire, Elegant Euphonia, and Shining Honeycreeper.

Natural lagoons and other wetlands, mainly in the pine savanna zone, are augmented locally by aquaculture farms, and together these habitats support good numbers of wading birds, including a breeding population of Jabirus, plus migrant waterfowl and shorebirds. Local concentrations of wading birds toward the end of the dry season can be spectacular.

Physical geography of Belize.

Especially in the north, increasing areas have been cleared of forest to support agriculture. The resultant farmland and associated rural habitats have promoted the spread of several open-country species in Belize, such as Crested Caracara, White-winged Dove, *Columbina* ground doves, Lesser Goldfinch, Bronzed Cowbird, and Eastern Meadowlark. The main terrestrial crops are sugar cane, bananas, and citrus, while pasture for cattle also covers increasingly large areas.

Much of Belize's coastline is fringed by mangroves, with relatively small areas of intertidal mudflats. A few river mouths can attract gatherings of shorebirds, gulls, and other waterbirds, notably the Belize River in Belize City and North Stann Creek in Dangriga.

The numerous inshore and offshore cayes add a distinctive Caribbean flavor to Belize's avifauna, with populations of White-crowned Pigeon, Caribbean Elaenia, and Smooth-billed Ani. Ambergris Caye, on the Mexican border, is actually a narrow peninsula, not strictly a caye, which explains its more diverse vegetation along with the relatively large number of mainland species found there compared with true island cayes.

Offshore cayes host a large breeding colony of Red-footed Booby (on Half Moon Caye) and two colonies of Magnificent Frigatebird, but former colonies of Atlantic Black Noddy, Common Brown Noddy, and Sooty Tern have been mostly or entirely wiped out by human disturbance. Relict breeding populations of Western Bridled and Roseate Terns seem likely to go the same way before long. Caribbean waters are poor for marine birds, home mainly to some boobies and terns, plus the occasional tropicbird, migrant jaeger, and storm-blown shearwater.

Rainforest understory, San Felipe Hills, Toledo. 23 December 2010. © *H. Lee Jones.*

Beautiful clear rivers and streams drain the Mountain Pine Ridge, here Privassion Creek, Cayo. 3 June 2020.
© *H. Lee Jones.*

Lowland pine savanna near Red Bank, Stann Creek. 2 July 2021. © *H. Lee Jones.*

Seasonally wet meadows are often intermixed with lowland pine savanna. Big Dry Creek meadow, Toledo. 8 March 2012. © *H. Lee Jones.*

Gallery forest along the Swasey River near Roseville, Toledo. 23 January 2018. © *H. Lee Jones.*

Stands of spiny bamboo occur locally, mainly along lowland rivers, and provide habitat for the very local Blue Seedeater. Monkey Bay Wildlife Sanctuary, Belize. 27 July 2021. © *H. Lee Jones.*

The open pine forests of Mountain Pine Ridge, Cayo, host several species not found in the lowland pine savannas. 10 January 2009. © *H. Lee Jones.*

Numerous wetlands and lagoons dot the Belize lowlands and are home to a good variety of waterbirds. Aguacaliente Lagoon, Toledo. 21 March 2021. © *H. Lee Jones.*

Increasing aquaculture, mainly in the coastal belt, provides habitat for numerous waterbirds. Aqua Mar Shrimp Farm, Toledo, 22 November 2012. © H. Lee Jones.

As water levels drop, aquaculture ponds can be attractive to wading birds. Caribbean Shrimp Farm, Belize. 27 July 2021. © H. Lee Jones.

Increasing areas of pastureland are facilitating the spread of many open-country species. Spanish Lookout, Cayo, 6 July 2021. © H. Lee Jones.

Rice farms also facilitate the spread of open country birds as well as providing seasonal habitat for crakes, migrant shorebirds, and other waterbirds. Toledo. 6 October 2010. © H. Lee Jones.

Red mangroves grow in areas permanently inundated by salt water, both along the mainland coast and on the cayes, here on Long Caye. 31 October 2015. © H. Lee Jones.

Fine white sand beaches, crystal clear waters, and swaying coconut palms typify many cayes. South Water Caye, Stann Creek. 9 March 2013. © H. Lee Jones.

TAXONOMY AND NAMES

Taxonomy is the science of classifying and naming things. For better or worse we are living in exciting if frustrating taxonomic times, as knowledge of avian taxonomy grows in leaps and bounds. Species are being shifted from one family to another, and families are being moved around to reflect their relatedness, despite the fact that it is inherently impossible to portray the multi-dimensional process of evolution in a linear list. One of the most heavily affected categories in this molecular revolution is that of genus, and the changes there are almost impossible to keep up with. Species, on the other hand, that elusive taxonomic level of most interest to birders, are being relatively ignored in all this higher-level work, where the focus is shifting toward genetic lineages, which may or may not reflect the more traditional concept of biological species.

Taxonomy and Species

For professional ornithologists, let alone birders, it can be challenging at best to keep pace with changing taxonomy and new names. Moreover, the advent of multiple bodies producing world lists—each with its own differing view of how species should be defined—plus the inconsistencies of what is treated as a species, or isn't, adds to the challenge. For example, Plain Wren *Thryothorus modestus* of Jones (2003) had its genus changed to *Cantorchilus* following a molecular study, and it has since been split by some authors into three species, including the northern member of the complex found in Belize, Cabanis's Wren *Cantorchilus modestus*, which at first glance is not immediately connected to the Plain Wren of old.

If it helps, when thinking about different species concepts, you might consider that the important word is *concept*, not *species*. That is, regardless of how we define a species, it basically comes down to a matter of opinion, with no right or wrong (Howell 2021).

For this guide, we have tried to address the species status of Belizean birds in terms of related taxa in North America, Central America, and in some cases South America or Eurasia when we have relevant experience. The ability to travel worldwide and observe birds, combined with literature and museum work, including sound analysis, has enabled us to offer opinions about many taxa. We evaluated species status on a case-by-case basis and with reference to how closely related taxa are treated. Our taxonomic review of Belizean birds revealed around 70 taxa that could represent 'new' species but which are not widely recognized as such, ranging from painfully obvious and in clear need of splitting to weakly differentiated and poorly known, as explained in Appendix B.

Ideally, a species' English name should be simple, informative, and preferably memorable. 'Golf Foxtrot Lima' we hear you respond to this sentiment. New names proposed here have for the most part attempted to disrupt the status quo as little as possible; most reflect geographic or plumage features, and a few commemorate persons who described the species or who have made major contributions to neotropical ornithology.

Many if not all the splits we adopt or suggest here will undoubtedly become 'official' in the future. We hope that drawing attention to them acts as a laxative on the taxonomic constipation manifested by some committees and speeds the rate at which ignorance and inertia fall victim to reality. If nothing else, our insights may help humans more meaningfully catalogue the burgeoning environmental 'anthropogenocide' being inflicted on our planet.

And for birders who just want a simple answer to know what they can count on their lists? Well, there is none. We make no apologies for the realistic if sometimes ambiguous course we have adopted because, as the relationship page on some websites might say: "It's complicated." Birders who keep lists can choose one of several options to follow, although it's a bit like different religions. The thinking person realizes sooner or later that these different checklists, or species concepts, can't all be right, but often there's some comfort, or convenience, in following one or the other.

SPECIES
ACCOUNTS

DUCKS (ANATIDAE; 12+ SPECIES) Familiar worldwide family usually associated with water. Ages differ slightly to distinctly; most species attain adult appearance in 1st year. Sexes similar in whistling ducks, different in other ducks.

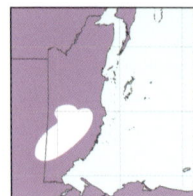

*BLACK-BELLIED WHISTLING DUCK *Dendrocygna autumnalis* 46–51cm. Distinctive, long-legged, boldly patterned duck with coral-red bill. Found in varied wetland habitats, from lakes and roadside ditches to flooded fields, mangroves. Feeds mostly at night, roosting by day in vegetation, occasionally in trees; feeds by dabbling and up-ending. Often in flocks, locally in 100s. In flight, note bold white upperwing stripe, dark underwing. Juv. duller overall with grayish bill and legs, ghosting of adult pattern; like adult within a few months. SOUNDS: High piping whistles, usually in short series of notes, mainly given in flight. STATUS: Fairly common to locally/seasonally common. (Tropical Americas.)

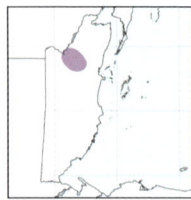

FULVOUS WHISTLING DUCK *Dendrocygna bicolor* 48–54cm. Distinctive, long-legged tawny duck of varied wetland habitats; associates readily with flocks of Black-bellied Whistling Ducks. In flight, note solidly dark wings, white U at base of tail. Juv. slightly paler, more weakly marked; like adult within a few months. Cf. rusty-stained Northern Pintail. SOUNDS: Nasal, slightly hoarse, usually 2-syllable whistles, distinct from piping and twittering of Black-bellied. STATUS: Uncommon to locally/seasonally common in north, mainly Crooked Tree area, with largest numbers late Feb–Aug; very rare and irregular elsewhere, including south. (Tropical Americas, Africa, Asia.)

MUSCOVY DUCK *Cairina moschata* 66–87cm, male>female. Large, distinctive, very heavy-bodied duck of fresh and brackish wetlands with cover; perches readily in trees. Widely domesticated, but wild birds usually wary, seen mainly in flight. Usually seen as singles or small groups, often not associating with other ducks. Feeds by dabbling, often while wading in shallows. Male appreciably larger than female, obvious when together; juv. dark overall, usually with white spot on wing; attains white forewings within 1st year. In some areas, wild and feral populations may mix: signs of impurity include an extensively pink face (vs. mostly dark on wild birds) and mixed black-and-white greater wing coverts (vs. solidly white). SOUNDS: Low quacks and hisses, rarely heard. STATUS: Uncommon to rare (where hunted) to locally fairly common (in more remote and protected areas). (Tropical Americas.)

DABBLING DUCKS (6+ species). All are winter migrants to Belize. Best identified by overall size and shape, wing patterns in flight; also note bill shape and color, leg color. Takeoff typically direct from water, without a running start, and often vocalize when disturbed.

NORTHERN PINTAIL *Anas acuta* 51–58cm. Winter migrant to varied wetland habitats, often in association with other ducks. Elegant and long-necked with tapered tail, slender blue-gray bill, grayish legs. Adult male distinctive (imm. resembles adult by early winter); other plumages paler and grayer than most dabbling ducks (but can be stained rusty in fall–winter, cf. Fulvous Whistling Duck), best identified by overall shape, bill shape and color, wing pattern: underwings dusky overall, upperwing has broad white trailing edge to speculum. SOUNDS: Male has high rolled *wirrrh*; female a slightly reedy quacking *kwerrk*. STATUS: Rare and irregular, mainly Nov–Mar. (Holarctic.)

AMERICAN WIGEON *Mareca americana* 43–51cm. Winter migrant to varied wetland habitats; often feeds by grazing on lakeshores, in fields. Medium-large and fairly stocky, with small gray bill tipped black, pointed tail, gray legs; white forewing of adult male striking in flight. Adult male distinctive (imm. resembles adult by early winter); female told by small gray bill, overall ruddy plumage with grayer head and neck, rather plain breast and sides. SOUNDS: Male has high breathy whistle, *whiih*, often doubled or in short series; female a low grunting quack. STATUS: Uncommon to rare and local mid-Oct to early Apr, mainly in north. (N America.)

BLACK-BELLIED WHISTLING DUCK

juv.

FULVOUS WHISTLING DUCK

imm.

MUSCOVY DUCK

females

domestic and feral birds variable

male

NORTHERN PINTAIL

females

male (Oct–Jun)

AMERICAN WIGEON

females

males (Oct–Jun)

BLUE-WINGED TEAL *Spatula discors* 38–41cm. Commonest, most widespread migrant duck in Belize, found in varied wetland habitats, especially rice fields; rarely on coastal waters. Often in flocks; flight frequently fast and twisting. Adult male distinctive (imm. resembles adult by mid-winter), other plumages cold-toned with whitish spot at base of bill, variable dark eyestripe. Cf. rare Cinnamon and Green-winged Teals. SOUNDS: Male has slightly reedy, piping *pseep*; female a low quack. STATUS: Fairly common to common Sep–Apr, a few from Aug and into May. (N America.)

NORTHERN SHOVELER *Spatula clypeata* 43–51cm. Winter migrant to varied wetland habitats, where feeds mainly by dabbling. Medium-large and stocky with big head, long spatulate bill, orange legs. Adult male distinctive; female told by big bill (longer than head); also note white tail sides. Imm. male resembles adult by late winter; many winter males messy, with whitish crescent forward of eyes. SOUNDS: Male gives low muffled grunts; female a quiet quack. STATUS: Uncommon to rare and local Oct–Mar, a few sometimes from mid-Sep and into early May; mainly in north. (Holarctic.)

CINNAMON TEAL *Spatula cyanoptera* 38–41cm. Rare winter migrant, usually found with flocks of Blue-winged Teal. Adult male distinctive (imm. resembles adult by mid-winter), other plumages similar to Blue-winged Teal but plumage warmer-toned (warm brown vs. gray-brown, but beware staining), with plainer face (lacking Blue-winged's distinct dark eyestripe, white spot at base of bill), bill slightly larger and more spatulate. Cf. female Northern Shoveler, Green-winged Teal. Mostly silent. STATUS: Rare to very rare and irregular mid-Sep to mid-Apr, mainly in north. (Americas.)

GREEN-WINGED TEAL *Anas [crecca] carolinensis* 35–38cm. Rare winter migrant to varied wetland habitats; often dabbles in muddy shallows. Adult male distinctive (imm. resembles adult by mid-winter); other plumages told by small size, pale wedge at sides of tail base, dull legs, dark upperwing with buff leading edge to speculum. Cf. Blue-winged Teal. SOUNDS: Male has high, reedy, piping *kriik*; female a low rough quack, vaguely snipe-like. STATUS: Rare to very rare and irregular Dec–Mar, mainly in north. (N America.)

SCAUPS (GENUS *AYTHYA*) (2+ species). Diving ducks that occur as winter migrants; like dabbling ducks, generally declining in Belize as birds remain farther north in winter. Best identified by overall size and shape, head and bill patterns, wing patterns in flight. Unlike dabbling ducks, patter across water to take off. Mostly silent in winter and spend much time sleeping.

LESSER SCAUP *Aythya affinis* 38–43cm. Winter migrant to wetland habitats with open water, including coastal waters; locally in flocks. Note slightly peaked hindcrown (head shape changes appreciably when diving; best appreciated when resting), broad white wingstripe mainly on secondaries. Among regularly occurring ducks, male distinctive (head sheen often purplish, but can be green), female told by big white patch at base of bill, wing pattern. Cf. Ring-necked Duck. Imm. male attains adult appearance by 1st spring, often looks like messy adult male in 1st winter. STATUS: Fairly common to common locally Nov–Mar in north (a few from mid-Oct and into Apr or later), rare to uncommon and irregular elsewhere. (N America.)

RING-NECKED DUCK *Aythya collaris* 41–46cm. Winter migrant to freshwater habitats, often with wooded edges; at times in small groups, and may associate with other ducks, especially Lesser Scaup. Note strongly peaked hindcrown, slender bill with white subterminal band, relatively long tail; in flight shows broad, poorly contrasting grayish wingstripe. Male distinctive, with black back; female best told by head shape, grayish head sides with narrow white spectacles, diffuse whitish patch at base of bill, wing pattern; cf. Lesser Scaup. Imm. male attains adult appearance by 1st spring, often looks like messy adult male in 1st winter. STATUS: Uncommon to rare and local late Oct–early Apr, mainly in north. (N America.)

BLUE-WINGED TEAL

females

male (Oct–Jun)

male (Jun–Mar)

females

NORTHERN SHOVELER

females

male (Oct–Jun)

CINNAMON TEAL

females

females

GREEN-WINGED TEAL

male (Oct–Jun)

male (Oct–Jun)

LESSER SCAUP

females

male (Oct–Jun)

RING-NECKED DUCK

females

male (Oct–Jun)

MASKED DUCK *Nomonyx dominicus* 33–38cm. Rare denizen of freshwater wetlands with emergent vegetation, from lake edges and small ponds to roadside ditches. May be found as singles or small groups, usually not associated with other species. Often well hidden in vegetation, at least during day; emerges at night on more open water to feed. Feeds by diving; rarely seen in flight (mainly dawn and dusk) but can spring into air, like teal. Breeding male distinctive; most records are of female-plumaged birds, which have distinctive buff face cut by 2 horizontal dark stripes (cf. young Black-bellied Whistling Ducks, which may be unattended by parents). Usually silent. **STATUS:** Scarce to rare and sporadic nonbr. visitor, with scattered records mainly Dec–Mar in north; possible almost anywhere. (Tropical Americas.)

GREBES (PODICIPEDIDAE; 2+ SPECIES) Cosmopolitan family of small to fairly large, essentially tailless, diving waterbirds; rarely seen in flight. Ages differ; like adult in 1st year; sexes similar, but male averages bigger bill.

LEAST GREBE *Tachybaptus dominicus* 21–24cm. Small dark grebe of fresh and brackish habitats, from weedy ponds and roadside ditches to lakes and mangroves; favors areas with emergent vegetation for cover, but sometimes on open water. At times associates with appreciably bulkier Pied-billed Grebe. Note slender dark bill, golden eyes; broad white wingstripe rarely seen. **SOUNDS:** Rapid, slightly pulsating, purring trill, recalling Ruddy Crake but slightly lower, burrier; also a quacking *kwrek*, and bleating *eirhk!* **STATUS:** Uncommon to fairly common, but often inconspicuous and local. (Tropical Americas.)

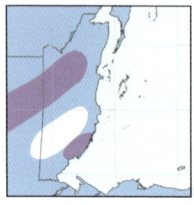

PIED-BILLED GREBE *Podilymbus podiceps* 28–33cm. Distinctive, chunky small grebe with stout pale bill. Found in varied habitats, from roadside ditches and small ponds to coastal lagoons, fish farms. At times associates loosely with other waterbirds, especially American Coot, Least Grebe. Narrow white trailing edge to wings rarely seen. **SOUNDS:** Complex 'song' a variable series of hollow clucks, coos, and grunts (can be given at night); also a rapid-paced, bleating chatter, and single quiet clucks. **STATUS:** Uncommon and sporadic local breeder, mainly in north; migrants occur widely but in low density, mainly Oct–Mar. (Americas.)

FINFOOTS (HELIORNITHIDAE; 1 SPECIES) Distinctive, small pantropical family. Despite the name, Sungrebes favor shady areas, are not closely related to grebes, and do not dive to feed. Ages/sexes differ slightly (male cares for young); attains adult appearance in 1st year.

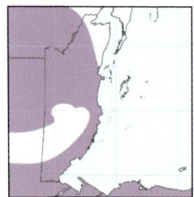

SUNGREBE *Heliornis fulica* 27–29cm. Distinctive but unobtrusive small swimming bird of unpolluted, typically slow-moving waterways with overhanging vegetation, including mangroves; can disperse to isolated lakes and ponds. Nothing really similar in Belize. Found as singles or pairs, usually swimming near shady cover. Picks for food on emergent and hanging vegetation and makes brief skittering dashes to snatch food near water surface. Flies readily when disturbed, pattering along the surface for takeoff; flight strong and low over water, when broad rounded tail conspicuous. Roosts on branches over water. Cheeks tawny on adult female, white on male; female bill and eyering flush red in breeding season. Juv. has whitish cheeks (tinged buff on female), duller bill than adult, pinkish-banded feet (yellowish-banded on adult). **SOUNDS:** Often quiet, but at times utters a sharp clucking *wek!* and (in territorial interactions?) a short series of (usually 4) barking clucks, *kwek! kwek! kwek! kw'eh*. 'Song' a series of far-carrying, hollow hoots, *ook, ook, …*, usually 1 about every 4 secs, sometimes faster paced. **STATUS:** Uncommon and local. (Tropical Americas.)

MASKED DUCK

breeding male
(year-round)

females

nonbr. male
(year-round)

duckling of
Black-bellied
Whistling Duck

LEAST GREBE

juv. (year-round)

nonbr.
(year-round)

breeding
(year-round)

PIED-BILLED GREBE

nonbr.
(year-round)

juv. (year-round)

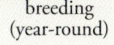

breeding
(year-round)

SUNGREBE

breeding
female

male

CORMORANTS (PHALACROCORACIDAE; 2 SPECIES) Worldwide family of rather large diving birds with long, hook-tipped bills, fully webbed feet. Ages differ, sexes similar; adult appearance attained in about 3 years; 2nd-year resemble dull adult. Often silent, but guttural grunts uttered in interactions and when nesting.

DOUBLE-CRESTED CORMORANT *Nannopterum (Phalacrocorax) auritum* 71–81cm. Large, heavy-bodied cormorant of coastal habitats and cayes; rarely inland on freshwater. Roosts on trees, pilings, sandbars, often alongside Neotropic Cormorant, which is smaller, more lightly built, with relatively longer tail; note extensive bright yellow facial skin of Double-crested (vs. dull lores on Neotropic); beware, face can be dull on juv. Double-crested, but gape skin rounded vs. more angular on Neotropic. Breeding adult attains bushy eyebrow crests; imm. variable, some dark brown overall, others fade to whitish on underparts. STATUS: Fairly common to common on coast and cayes; increasing overall and spreading south. Scarce and sporadic but increasing visitor inland in north. (N America.)

NEOTROPIC CORMORANT *Nannopterum (Phalacrocorax) brasilianum* 64–70cm. The common cormorant throughout most of Belize. Found in varied wetland habitats, from ponds and rivers to mangroves and inshore coastal waters. Habits much like Double-crested Cormorant (which see for differences); both species often perch with wings outstretched. Flight often a little quicker, more maneuverable than Double-crested, which, with rather slim shape, can suggest dark ibises. Plumages parallel Double-crested (but lacks crests; breeding adult attains variable white streaks on head and neck); imm. can fade to mostly whitish below. Cf. Anhinga, which often occurs alongside in the same habitats. STATUS: Common to fairly common almost throughout; recorded occasionally on inshore cayes but not offshore cayes, where Double-crested occurs. (Americas.)

ANHINGAS (ANHINGIDAE; 1 SPECIES) Small pantropical family of long-necked, very long-tailed diving waterbirds with sharply pointed bills, fully webbed feet. Ages/sexes differ; adult appearance attained in 2–3 years. Wing molt synchronous, unlike gradual wing molt of cormorants.

ANHINGA *Anhinga anhinga* 84–94cm, WS 109–119cm. Long-necked and long-tailed diving bird of varied wetland habitats, from small wooded ponds to lakes, rivers, mangroves on inshore cayes. Distinctive, but cf. Neotropic Cormorant, commonly in the same areas; note sharp pointed bill, white upperwing panels, and broad pale tail tip of Anhinga. Perches on trees over water, often with wings outstretched, and soars readily on thermals; soaring at a distance may suggest a kite or large falcon. Flies with deep strong wingbeats, alternated with glides on flattish wings, long tail usually spread slightly. Often swims with only neck above the surface, jerking along, hence a local name 'snakebird' (cormorants can also do this, however). Juv. has dirty buff head and neck, brown belly, duller wing panels than adult. SOUNDS: Rasping and creaky short croaks, often in slightly stuttering or descending series. STATUS: Uncommon to fairly common on mainland; occasional visitor to Ambergris Caye, mainly fall–winter. (Tropical Americas.)

DOUBLE-CRESTED CORMORANT

adults

breeding

imm.

adult

imm.

NEOTROPIC CORMORANT

imm.

adults

adults

imm.

ANHINGA

males

females

breeding colors

FRIGATEBIRDS (FREGATIDAE; 1 SPECIES) Very large but very light-bodied, mostly black seabirds with long crooked wings, deeply forked tails that can be held closed in a point. Ages/sexes differ; adult male has red throat pouch inflated in display; adult appearance attained in about 5–6 years.

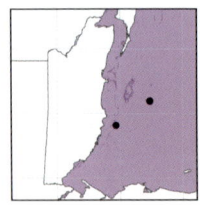

MAGNIFICENT FRIGATEBIRD *Fregata magnificens* 90–110cm, WS 200–240cm. Distinctive and conspicuous species of coastal and marine waters; often around fishing harbors, roosting on ship rigging and in mangroves. Does not alight on water; ranges inland a short distance to drink fresh water and splash-bathe at small lakes, ponds, rivers. Breeds locally on offshore cayes, building stick nests in mangroves. Flight easy and buoyant with slow deep wingbeats, frequent effortless soaring, often in kettles high overhead; plucks food from sea surface with long hooked bill and also pirates terns, boobies, other seabirds. No similar species in Belize. Adult male wholly glossy black; female has black head, white chest; juv. has white head and body; complex age/sex plumage progression to adult plumage. SOUNDS: Soft wheezy warbling and bill rattling, mostly in display. STATUS: Fairly common to common along coast and over marine waters; breeding colonies on Man O'War Caye and Half Moon Caye. Rare and sporadic wanderer inland. (Tropical Americas and Galapagos.)

PELICANS (PELECANIDAE; 2 SPECIES) Small worldwide family of very large, heavy-bodied aquatic birds with long bills and distensible throat pouches. Difficult to misidentify. Ages differ, sexes similar; attain adult appearance in about 3 years. Some seasonal change in adult appearance. Adults mostly silent, rarely uttering grunts and hisses; begging juvs. can be noisy at nests.

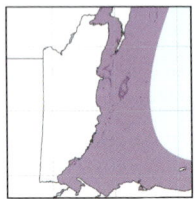

BROWN PELICAN *Pelecanus occidentalis* 112–137cm, WS 190–255cm. Essentially unmistakable large dark waterbird of coastal and inshore waters, river mouths, adjacent lagoons, mangroves; rarely inland; nests locally on cayes. Often rests on sandbars and beaches with gulls, terns, cormorants, other waterbirds, also perches on pilings, boats, mangroves. Singles and lines fly low over the waves, gliding easily between bouts of measured flapping; at times sails and soars high overhead. Feeds by plunge-diving, twisting abruptly on entry into the water. Adult silvery gray above, dark below, with white head and neck in nonbr.; hindneck molts to dark brown in breeding plumage, eye becomes whitish. 1st-year dark brown overall with white belly, broad white median stripe on dark underwing; 2nd-year like messy adult. STATUS: Fairly common to common along coast and on cayes, breeding locally; rare and sporadic wanderer inland. (Americas.)

AMERICAN WHITE PELICAN *Pelecanus erythrorynchos* 145–165cm, WS 240–290cm. Huge, mostly white migrant waterbird of wetlands, coastal lagoons; very rarely on the ocean or even on adjacent beaches. Essentially unmistakable, but at long range in flight cf. Wood Stork, which tends to soar in less synchronized kettles. Often in groups, locally of 100s; feeds while swimming and submerging its bill, not by diving. Often seen soaring and circling, at times in kettles with other waterbirds and vultures. 1st-year has paler, more pinkish face, bill, and legs than adult, extensive black on inner secondaries and greater coverts, faint dusky wash to upperwing coverts. 2nd/3rd-year has brighter, more orangey face, bill, and legs, less black on upperwings than 1st-year, often develops dark mottling on upperwing coverts. STATUS: Uncommon to fairly common but local winter migrant, mainly Nov–Apr in north, but small numbers might appear anywhere; occasionally a few occur through the summer; rare wanderer to Ambergris Caye. (Breeds N America to n. Mexico, winters to Cen America.)

Magnificent Frigatebird

adult males

1st-year

adult female

Brown Pelican

adults

1st-years

nonbr.

breeding

American White Pelican

nonbr.

BOOBIES (SULIDAE; 3 SPECIES) Small worldwide family of large, streamlined, plunge-diving oceanic birds. Ages differ, sexes similar or differ in face/bill colors, voice; attain adult appearance in 2–4 years. Vocal mostly on breeding grounds but can be heard in feeding interactions.

ATLANTIC BROWN BOOBY *Sula leucogaster* 68–75cm, WS 135–153cm. Most frequently seen booby in most of Belize. Usually singles, occasionally small groups, feeding or rafting on water, at times with other seabirds; perches on buoys, pilings. Slightly larger and bulkier than Red-footed Booby, with heavier flight, less crooked wings; imm. body darker than whitish underwing coverts (reverse of imm. Red-footed) and upperparts solidly brown, lacking white tail tip often shown by imm. Red-footed. Cf. imm. Masked Booby, which has stout yellowish bill, variable whitish hindcollar, soon develops white back patches. Adult male has pale ivory bill, female bill pinkish. SOUNDS: Male gives high wheezy whistles, female gruff brays. STATUS: Uncommon year-round over offshore waters and around smaller cayes; occasionally seen from mainland, especially in windy weather. (Tropical Atlantic.)

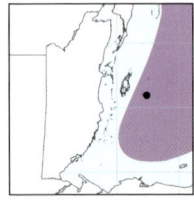

RED-FOOTED BOOBY *Sula sula* 66–76cm, WS 130–150cm. Mainly offshore except when breeding at Half Moon Caye; exceptionally seen from mainland. Occurs at sea with mixed-species feeding flocks of other boobies, terns; often curious around boats and ships, roosts on rigging. Smallest, most lightly built booby with highly variable plumage. Diagnostic bright red feet develop in 2nd year, pinkish on juv. (rarely pinkish on juv. Atlantic Brown Booby). Note crooked wings, long tail, maneuverable flight. Imm. from Atlantic Brown Booby by structure, paler body contrasting with dark underwings; imm. bill often pinkish with dark tip and tail usually has white tip. Adults in Belize are white-tailed, with white morph much commoner than brown; also note pink-and-blue face, pale bluish bill. Cf. adult of larger and bulkier Masked Booby, which has more black on wings, black tail, stout yellowish bill. SOUNDS: Both sexes utter low guttural brays and chatters. STATUS: Common breeder (mainly Nov–Jul) on Half Moon Caye, ranging over offshore waters; very rare around inshore cayes and on mainland, mainly in association with tropical storms. (Tropical Oceans worldwide.)

MASKED BOOBY *Sula dactylatra* 73–81cm, WS 150–170cm. Rare, mainly offshore but at times roosts on piers or boat docks and can be confiding. At sea likely to be alone or with feeding flocks of other boobies, terns. All ages have stout yellowish bill, yellowish legs and feet, dark face, extensively white underwings, black tail. 1st-year has broad white neck collar, often some white on back and rump. Cf. imm. with adult Atlantic Brown Booby (solidly dark upperparts), adult with white morph Red-footed Booby (white tail, bluish bill). SOUNDS: Male gives high wheezy whistles, female gruff brays. STATUS: Rare offshore visitor, possible year-round; very rare on mainland and on inshore cayes, mainly Aug–Nov in association with tropical storms. (Tropical Oceans worldwide.)

ATLANTIC BROWN BOOBY

male

1st-years

adult females

RED-FOOTED BOOBY

older imm.
white morph

1st-years (all morphs similar)

adults

adults

adults

MASKED BOOBY

juv.

adult

juvs.

adults

JAEGERS (STERCORARIDAE; 2+ SPECIES) Small family of rather gull-like oceanic birds that feed mainly by pirating other birds, mainly gulls, terns. Ages differ, sexes similar with females averaging larger. Adult plumage attained in about 3 years. Known as skuas in Old World.

PARASITIC JAEGER (ARCTIC SKUA) *Stercorarius parasiticus* 40–44cm (+ 6–10cm adult tail projections). Most likely jaeger to be seen from shore; chases mainly terns, smaller gulls. Flight strong and direct, suggesting a falcon; chases often persistent and aerobatic. Immature and nonbreeding plumages highly variable, but note relatively small head, slender bill, sharply pointed tail projections, crescent of white primary shafts on upperwing. All ages can be dark overall with reduced white wing flashes; adult dark morph fairly common. Juv. often relatively rusty-toned, unlike colder-toned juv. Pomarine, with streaked nape (mottled on Pomarine). 1st-year has heavily barred underwings, 2nd-year has reduced barring on underwing coverts, 3rd-year and older typically have solidly dark underwing coverts. **STATUS:** Scarce nonbr. migrant offshore, mainly Aug–Apr, most likely in late fall and spring; seen occasionally from mainland, especially off Belize City. (Breeds n. Eurasia and N America, winters widely at sea.)

POMARINE JAEGER (SKUA) *Stercorarius pomarinus* 44–51cm (+ 6–11cm adult tail projections). Largest jaeger, unlikely to be seen from shore; chases mainly gulls, larger terns. Direct flight rather heavy and steady, with powerful wingbeats; chases rarely prolonged and aerobatic. Sometimes scavenges at fishing boats. Immature and nonbreeding plumages highly variable, but note relatively big head and bill, broad wings, and broad, blunt-tipped tail projections. All ages can be dark overall with reduced white wing flashes; adult dark morph uncommon. Ageing as in Parasitic Jaeger. **STATUS:** Scarce nonbr. migrant offshore, few confirmed records; most likely in late fall–winter and spring. (Breeds n. Eurasia and N America, winters widely at sea.)

GULLS (LARIDAE; 5+ SPECIES) Widespread family of familiar web-footed birds often found near water. Ages differ, sexes similar but males average larger, bigger-billed. Adult appearance attained in 2 years for smaller species (hence, 2-year gulls), up to 4 years or longer in large species (4-year gulls). Seasonal variation mainly in head pattern, bill color and pattern. Varied crowing and mewing calls given mainly in displays and feeding interactions.

LESSER BLACK-BACKED GULL *Larus fuscus* 53–62cm. Rare winter migrant. A relatively long-winged, large 4-year gull, most likely to be found as singles on beaches, at harbors, river mouths, coastal lagoons; associates readily with other gulls. Averages smaller and more lightly built than Smithsonian Gull, but male appreciably larger than female. Adult relatively distinctive: note size and structure, yellowish legs, slaty-gray upperparts (palest birds from W Europe similar in tone to Laughing Gull, others—from N Europe—appreciably darker); nonbr. has extensively dusky head and neck streaking. 1st-year mottled brown overall, averages darker than Smithsonian Gull but often with whitish head and breast, best identified by relatively slender blackish bill, long narrow wings without distinct pale panel on inner primaries of Smithsonian, contrasting black/white tail pattern. 2nd- and 3rd-years highly variable: note long wings, bill shape, slaty-gray back plumage; legs can be pinkish overall into 3rd winter. **STATUS:** Rare to very rare but increasing late Oct–Mar, mainly on n. coast and cayes; first recorded Belize in 2011. (W Eurasia; spread to New World and increasing, mainly since 1970s.)

PARASITIC JAEGER

juv.
(Aug–Mar)

nonbr.
(Aug–Mar)

dark morph

breeding
(Mar–Oct)

POMARINE JAEGER

juv.
(Sep–Mar)

nonbr.
(Aug–Mar)

dark morph

breeding
(Mar–Oct)

LESSER BLACK-BACKED GULL

N Europe

1st-years
(Sep–Jul)

adult nonbreeding
(Aug–Mar)

W Europe

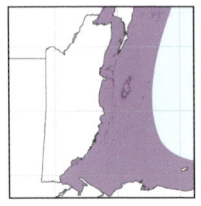

LAUGHING GULL *Leucophaeus atricilla* 38–43cm. Commonest gull in Belize, mainly coastal. Medium-size, long-winged 3-year gull with dark bill and legs. Found on beaches, at river mouths, coastal lagoons, dumps, over inshore waters; rarely inland. Locally in flocks of 100s, often with terns, other gulls. Fairly distinctive, but cf. Franklin's Gull. Note long wings, relatively heavy, often slightly droop-tipped bill. Nonbr./imm. has distinctive smudged dusky mask through eyes. Juv. dark brown overall with scaly pale edgings above; soon attains gray back, whiter head and underparts. 2nd-year like dull adult with more black in wing-tip, sometimes black in tail. **SOUNDS:** Varied, nasal laughing and yelping cries. **STATUS:** Common to fairly common on coast and cayes, more local in summer (uncommon and very local breeder on some s. cayes); rare and sporadic wanderer inland. (Breeds N America to Caribbean, winters to S America.)

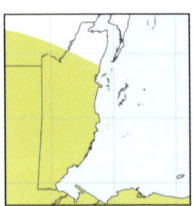

FRANKLIN'S GULL *Leucophaeus pipixcan* 35–38cm. Scarce transient, mainly coastal. Medium-small 3-year gull with dark bill and legs. Singles or rarely flocks, usually associating with Laughing Gulls on beaches, at river mouths, coastal lagoons, over inshore waters; also possible high overhead inland during spring migration. Slightly smaller than Laughing Gull, with smaller bill, more rounded wing-tips, thicker white eye-arcs; spring adults often have strong pink blush to underparts (Laughing can have pale blush). Adult wing-tip boldly patterned black-and-white, unlike Laughing, and imm./nonbr. plumages have distinctive blackish half-hood. **SOUNDS:** Yelping calls higher, more mewing than Laughing Gull. **STATUS:** Scarce and sporadic transient, mainly Oct–Dec, Mar–May; very rare in winter. Peak numbers in early to mid-May, when exceptional flocks of 100s in 2017. Most records from coastal areas, but could occur anywhere. (Breeds N America, winters S America.)

RING-BILLED GULL *Larus delawarensis* 44–52cm. Medium-size 3-year gull with pale gray back. Found mainly as singles with other gulls on beaches, at river mouths, rarely over inshore waters. Adult distinctive, with yellow legs, pale eyes, neat black bill ring (no red on bill). 1st-year rather pale overall, with pinkish legs, whitish underwings, clean-cut black/white tail pattern, pale gray inner primaries; cf. older imm. of larger Smithsonian Gull. 2nd-year Ring-billed resembles duller adult with more black in wing-tip, often some black on tail, greenish-yellow legs. **STATUS:** Rare and local in coastal areas, mainly Sep–Apr at Ambergris Caye and Belize City; very rare at other seasons and inland. (Breeds N America, winters to Cen America.)

*SMITHSONIAN (AMERICAN HERRING) GULL *Larus [argentatus] smithsonianus* 56–67cm. The default species of large 4-year gull in Belize; other species occur as very rare visitors or vagrants. All ages have pink legs; male appreciably larger and bigger-billed than female. Found mainly as singles with other gulls at river mouths, coastal lagoons, beaches. Adult (rare in Belize) distinctive, with pink legs, black wing-tips, pale eyes, variable dusky streaking on head and neck in nonbr. 1st-year mottled brownish overall, cf. imm. of rare Lesser Black-backed Gull. 2nd- and 3rd-years highly variable; 2nd-year pattern resembles smaller 1st-year Ring-billed Gull but messier, lacks clean black/white tail pattern of Ring-billed. **STATUS:** Uncommon to rare and local mid-Oct to Apr, mainly in n. coastal areas; very rare at other seasons and inland. (Breeds N America, winters to Cen America.)

LAUGHING GULL

breeding (Jan–Aug)

1st-years
(Aug–Jun)

adults

nonbr.
(Jul–Mar)

FRANKLIN'S GULL

breeding (Feb–Jul)

adults

1st-years
(Sep–Mar)

nonbr.
(Jul–Mar)

breeding (Feb–Aug)

RING-BILLED GULL

adults

1st-years (Sep–Jul)

nonbr.
(Aug–Mar)

older imms.

breeding (Feb–Aug)

adults

1st-years (Sep–Jul)

SMITHSONIAN GULL

nonbr.
(Aug–Mar)

TERNS (LARIDAE; 12+ SPECIES) Worldwide group of waterbirds that resemble gulls but have pointed bills, shorter legs, and typically are smaller, more graceful, with forked tails. Unlike gulls, rarely alight on water, mostly feed by plunge-diving for small fish. Ages differ, sexes similar but males average larger, bigger-billed. Adult appearance attained in 2–3 years; imm. plumages typically resemble nonbr. adults. Seasonal variation mainly in head pattern.

GULL-BILLED TERN *Gelochelidon nilotica* 33–36cm. Medium-size, rather stocky tern of lagoons, beaches, lakes, wetlands; not over the ocean. Feeds by distinctively swooping down to snatch prey (crabs and such) from ground and shorelines, not by diving into water. Associates readily with other terns, gulls, skimmers. Distinctive: note habits, ghostly pale gray upperparts (no white rump), very tapered and swept-back wings, short tail, thick black bill, relatively long legs. Cf. nonbr./imm. Forster's Tern. SOUNDS: Nasal laughing and mellow barking calls, mainly in flight, such as *ku-wek* and *ket-e-wek*. STATUS: Uncommon to locally fairly common Aug–May, very rare in summer; most numerous Feb–Apr at Crooked Tree. (Worldwide.)

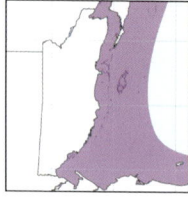

*SANDWICH TERN *Thalasseus sandvicensis* 34–36cm (+ 2.5cm adult tail streamers). Distinctive, medium-size tern of beaches, inshore waters, river mouths. Often in flocks with other terns, gulls, skimmers. Note shaggy crest, slender black bill tipped yellow on adult. Juv. bill can be yellowish with dark tip, soon becomes black for 1st year. Cf. Gull-billed Tern. SOUNDS: Grating, screechy and rasping calls, such as *krríik*, distinct from *Sterna* terns; 1st-year has high piping whistles. STATUS: Fairly common to common year-round along coast, over inshore waters, on cayes, most numerous in winter; local breeder on some smaller cayes. (Breeds N Atlantic, winters to S Atlantic.)

ROYAL TERN *Thalasseus maxima* 43–48cm (+ 5cm adult tail streamers). Large, orange-billed tern of coastal habitats. Often in feeding and roosting flocks with other terns, gulls, skimmers. Slightly smaller and more lightly built than Caspian Tern, with narrower, more angled wings, and longer tail, shaggier cap, uniform orange bill (rarely orange-red); lacks solidly dark underside to wing-tip of Caspian; imm./nonbr. plumages have large 'bald' white forehead, unlike Caspian. Legs rarely orange, mainly on imms. SOUNDS: Adult has clucking *krehk* and laughing *kewh-eh*; also grating and screechy calls that may suggest Sandwich Tern; 1st-year has high piping whistles. STATUS: Fairly common to common Sep–Apr along coast, over inshore waters, on cayes; smaller numbers of nonbreeders remain locally in summer; very rare inland. (Americas.)

CASPIAN TERN *Hydroprogne caspia* 51–57cm. Largest tern in the world, a scarce but widespread migrant to coastal lagoons, wetlands, river mouths, beaches, very rarely over open ocean. Often rests with groups of other waterbirds, especially gulls, other terns, mainly as singles or small groups. Bulky and broad-winged, lacks long tail streamers. Wingbeats relatively shallow and gull-like. Note overall size and bulk, very stout red bill with black mark near tip, dark underside to primaries; densely black-streaked crown of nonbr./imm. Juv. has orange bill with dark near tip. Cf. Royal Tern. SOUNDS: Adult has deep throaty *rahrr*, quite distinct from other terns and may suggest a heron; 1st-year has high lisping whistles. STATUS: Uncommon to rare and local Oct–May, mainly along n. coast and locally inland, especially Crooked Tree in spring; a few nonbreeders may remain in summer. (Worldwide except S America.)

breeding
(Mar–Aug)

GULL-BILLED TERN

nonbr.

nonbr.
(Aug–Mar)

SANDWICH TERN

1st-year

breeding
(Feb–Jul)

nonbr.
(Jul–Feb)

ROYAL TERN

1st-year

breeding
(Mar–Jul)

nonbr.
(Jul–Mar)

CASPIAN TERN

breeding
(Mar–Sep)

1st-year

nonbr.
(Sep–Mar)

LEAST TERN *Sternula antillarum* 21–23cm. Tiny summer migrant tern of coastal waters, cayes, sandy beaches, river mouths; nests in colonies on sandy beaches. Tiny size distinctive—body barely larger than a Sanderling! Feeds mainly in coastal and inshore waters; often rests with other terns, smaller gulls, shorebirds. Flies with hurried deep wingbeats and plunge-dives steeply from moderate heights. **SOUNDS:** Varied, slightly sneezy and squeaky calls, typically a 2-syllable *chírit* and *kree-it*. **STATUS:** Fairly common but local and declining breeder Apr–Aug on a few cayes and along mainland coast; migrants occur more widely Aug–Oct, mid-Mar to May, when very rare inland. Not well known in winter, but may occur rarely offshore. (Breeds N America to Mexico, winters Mexico to n. S America.)

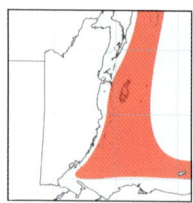

ROSEATE TERN *Sterna dougallii* 30–33cm (+ 4–5cm adult tail streamers). Lightly built summer migrant tern of inshore marine waters, cayes, rarely along mainland coast with other terns. Smaller-bodied and longer-tailed than Common Tern, with more slender bill; rather hurried wingbeats can suggest Least Tern. Breeding adult whiter overall than Common with dusky leading edge to outer primaries but no distinct dark trailing edge; tail white with very long streamers, lacks dark outer web to outermost feathers of Common. Rosy blush to underparts rarely noticeable; bill mostly black Apr–May, develops red base Jun–Aug. Nonbr. whiter than Common, without distinct dark shoulder bar. **SOUNDS:** Calls include scratchy 2-syllable *kirrik*, unlike Common Tern but reminiscent of Sandwich Tern. **STATUS:** Uncommon, very local, and declining breeder Apr–Aug on cayes; very rarely seen on mainland coast, mainly Sep–Dec. (Worldwide.)

COMMON TERN *Sterna hirundo* 29–32cm (+ 2.5cm adult tail streamers). Coastal and offshore nonbr. migrant, often resting on beaches, at river mouths, less often in coastal lagoons; feeds mainly over inshore marine waters. Singles or small flocks, often with other terns, gulls. On nonbr./imm. note partial black cap with white forecrown, blackish leading edge to wing (shows at rest as dark shoulder bar). Breeding adult has red bill with small dark tip, pale smoky wash to body. Post-juv. plumages often have dark wedge on trailing edge of primaries (can be indistinct in spring, and also shown by 2nd-year Forster's Tern). Wing molt occurs fall–winter, unlike Forster's. Cf. Forster's and Roseate Terns. **SOUNDS:** High sharp *kiik*, suggesting Long-billed Dowitcher. **STATUS:** Uncommon to seasonally fairly common on coast and cayes, mainly Aug–Nov, Apr–May; scarce in winter. (Breeds n. Eurasia and N America, winters to S Hemisphere.)

FORSTER'S TERN *Sterna forsteri* 32–36cm (+ 6–7.5cm adult tail streamers). Scarce winter migrant to wetlands, fish farms, coastal lagoons, beaches; rarely offshore. Likely to be found as singles, often resting with other terns, gulls, shorebirds. Slightly larger, bigger-billed than Common Tern. Nonbr./imm. plumages have distinctive black face mask, orange-red legs, cf. Gull-billed Tern. Breeding adult has silvery upperwings, whitish body, orange-red bill with extensive black tip; wing molt completes before winter. Cf. Common Tern. **SOUNDS:** Hard clipped *kik!* **STATUS:** Scarce and irregular Oct–Apr; very rare at other seasons. (Breeds N America, winters to Cen America.)

LEAST TERN

breeding
(Mar–Aug)

juvs.
(Jul–Sep)

1st-year

nonbr.
(Jul–Mar)

ROSEATE TERN

breeding
(Apr–Sep)

juvs.
(Jul–Aug)

near fledging

nonbr.
(Aug–Mar)

COMMON TERN

breeding
(Apr–Sep)

1st-year

nonbr.
(Sep–Mar)

FORSTER'S TERN

breeding
(Mar–Aug)

nonbr.

nonbr.
(Aug–Mar)

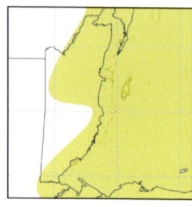

*AMERICAN BLACK TERN *Chlidonias [niger] surinamensis* 23–25cm. Transient migrant of marine waters, coastal lagoons; also inland at wetlands, shrimp farms. Mainly singles and small groups, associating readily with other terns, but flocks of 100s can occur offshore in fall. Flight slightly floppy, swooping to pick food from near water surface, not plunge-diving; rests on flotsam, sea turtles. Very small size, dusky gray upperparts, and dark spur on sides of breast distinctive; black-bodied full breeding plumage rarely seen in Belize. **SOUNDS:** Quiet piping whistles on occasion. **STATUS:** Uncommon to fairly common, mainly Jul–early Nov, Apr–early Jun; most numerous in fall offshore. (Breeds N America, winters Mexico to S America.)

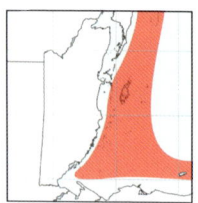

WESTERN BRIDLED TERN *Onychoprion [anaethetus] melanopterus* 33–36cm (+ 5–6.5cm adult tail streamers). Offshore waters; nests in small colonies. Usually singles or small loose groups, at times with feeding flocks of boobies, other terns. Flight buoyant and graceful, swooping to pick from surface rather than plunge-dive; often rests of flotsam. Adult has dark gray-brown upperparts, extensively white tail and underwings, long white brow, cf. Sooty Tern. Imm./nonbr. has pale edgings to upperparts, messier head pattern. **SOUNDS:** Braying and clucking calls, lower and more grating than Sooty Tern. **STATUS:** Uncommon to fairly common but declining local breeder on cayes, Mar–Aug, with migrants into Sep–Oct; very rare on mainland, mainly in association with tropical storms. (Tropical Americas, Atlantic.)

SOOTY TERN *Onychoprion fuscatus* 36–39cm (+ 6.5–7.5cm adult tail streamers). Offshore waters. Often in flocks, feeding over schooling fish with boobies, noddies; circles high when searching for food, soaring easily on fairly broad wings, unlike Western Bridled Tern. Adult clean black-and-white with white forehead patch, extensive dark on underside of primaries, cf. Western Bridled Tern. Juv. distinctive, dark sooty brown overall, spotted white to buff above, with contrasting whitish underwing coverts; cf. Common Brown Noddy. **SOUNDS:** Varied clucking calls, including *wed-a-wek*. **STATUS:** Uncommon to fairly common offshore Feb–Aug. Mostly extirpated as a breeding bird from cayes, courtesy of human disturbance, but may persist locally. Very rare on mainland, in association with tropical storms. (Pantropical.)

COMMON BROWN NODDY *Anous stolidus* 36–42cm. Offshore waters, usually as singles or small groups feeding with boobies, other terns. Feeding flight typically low to the water, swooping to pick from the surface; transiting flight low and direct, when can suggest a small jaeger. Rests on flotsam. Habits and uniform dark brown plumage (including underwings, cf. juv. Sooty Tern) distinctive; adult has whitish forecrown, juv. has narrow whitish bridle. **STATUS:** Scarce offshore, mainly Apr–Oct but reported in all months; very rare on mainland, in association with tropical storms. Formerly bred locally on cayes. (Pantropical.)

SKIMMERS (1 species) Small pantropical group treated as a distinct family or as a subfamily within gull and tern assemblage. Ages differ, attaining adult appearance in 1st year; sexes similar but male has appreciably larger bill.

*BLACK SKIMMER *Rynchops niger* 43–45.5cm, WS 115–123cm. Rare. Distinctive, angular, and boldly patterned waterbird that may be found at river mouths, sandbars, coastal lagoons. Singles or small groups rest with gulls and terns. Flocks typically fly in rather compact, wheeling formation, wingbeats mainly above body plane. Feeds in flight, mostly at night, by slicing elongated mandible through water surface and snapping shut on contact with food. Breeding plumage has solidly black hindneck; juv. has pale edgings to upperparts. **SOUNDS:** Nasal laughing and barking calls, *kruh* and *kwuk*; calls mainly in flight, including at night. **STATUS:** Rare and irregular Nov–Mar along coast, especially around Belize City and Dangriga; very rare and sporadic inland. (Americas.)

AMERICAN BLACK TERN

nonbr.
(Jul–Mar)

breeding
(Apr–Jul)

WESTERN BRIDLED TERN

breeding
(Feb–Aug)

juvs.
(Jun–Oct)

juvs.
(Jun–Nov)

SOOTY TERN

breeding
(Jan–Sep)

adult

1st-year

**COMMON
BROWN NODDY**

adult

breeding
(Mar–Aug)

BLACK SKIMMER

nonbr.
(Aug–Mar)

SHOREBIRDS (34+ SPECIES) For ID purposes there are 3 basic types of shorebirds: 3 unmistakable large species (stilt, avocet, oystercatcher; opposite); 7+ plovers (visual feeders, with stop-start feeding actions; pp. 50–53); and 23+ sandpipers (mainly tactile feeders, picking and probing as they walk along; pp. 54–63). Most favor open habitats, typically near water, and different species often associate together, which can help greatly with ID— compare size, shape, bill shape, and behavior of an unfamiliar species with other species you know; voice and habitat can also be useful.

STILTS AND AVOCETS (RECURVIROSTRIDAE; 2 SPECIES) Small
worldwide family of elegant, long-legged shorebirds found in warmer climates. Ages/sexes differ slightly, avocet has seasonal plumage changes; adult plumage attained in 2nd year.

BLACK-NECKED STILT *Himantopus mexicanus* 36–41cm. Essentially unmistakable, a visually elegant but vocally irritating large shorebird with improbably long, hot-pink legs. Breeds at lagoons, shrimp farms; ranges to varied wetland habitats, less frequent on open coast. Often in flocks, associating readily with other shorebirds, and often breeds in small colonies. Male has glossy black back and wings, breast often flushes pink on breeding birds; female and imm. back brownish; juv. has weaker dark head and neck pattern, whitish trailing edge to secondaries; like adult in 1st year. **SOUNDS:** Varied, often persistent yapping and clucking calls, especially when nesting, have earned the nickname 'Marsh Poodle;' sharp piping *kiip* suggests Long-billed Dowitcher. **STATUS:** Fairly common to locally common, especially in winter; more local and less numerous in summer, but increasing with expansion of shrimp farms. (N America to n. S America.)

AMERICAN AVOCET *Recurvirostra americana* 40–43cm. Rare nonbr. migrant. Elegant and essentially unmistakable shorebird with fine, upcurved bill, boldy pied plumage. May be found in varied wetland habitats, especially shrimp farms, river mouths, coastal lagoons. Singles or small groups associate readily with other shorebirds, especially stilts. Sexes similar, but male has straighter bill than female. Juv. has rusty-tinged neck, like adult nonbr. by winter. **SOUNDS:** Overslurred piping *kleéh*, singly or in series, at times persistently repeated; can recall oystercatchers. **STATUS:** Rare and irregular Oct–Mar, very rarely lingering into May–Jun. (Breeds N America to Mexico, winters to Cen America.)

OYSTERCATCHERS (HAEMATOPODIDAE; 1 SPECIES) Small worldwide
family of large stocky shorebirds with laterally compressed, bright orange-red bills, thick pink legs. Ages differ slightly, sexes similar; like adult in 2nd year.

AMERICAN OYSTERCATCHER *Haematopus palliatus* 40.5–45.5cm. Rare and local nonbr. migrant. A distinctive, large, boldly patterned shorebird of sandy beaches, river mouths, mudflats. Usually occurs in ones and twos, often associating with other waterbirds at high-tide roosts. 1st-year has duller eyes and legs, dark-tipped bill. **SOUNDS:** Loud piping and screaming calls, *Wheeh* and *h'wheek*, often run into shrill piping chatters, at times prolonged. **STATUS:** Rare to very rare and local in coastal areas, including inshore cayes, mainly Sep–Mar; most records from Dangriga. (Americas.)

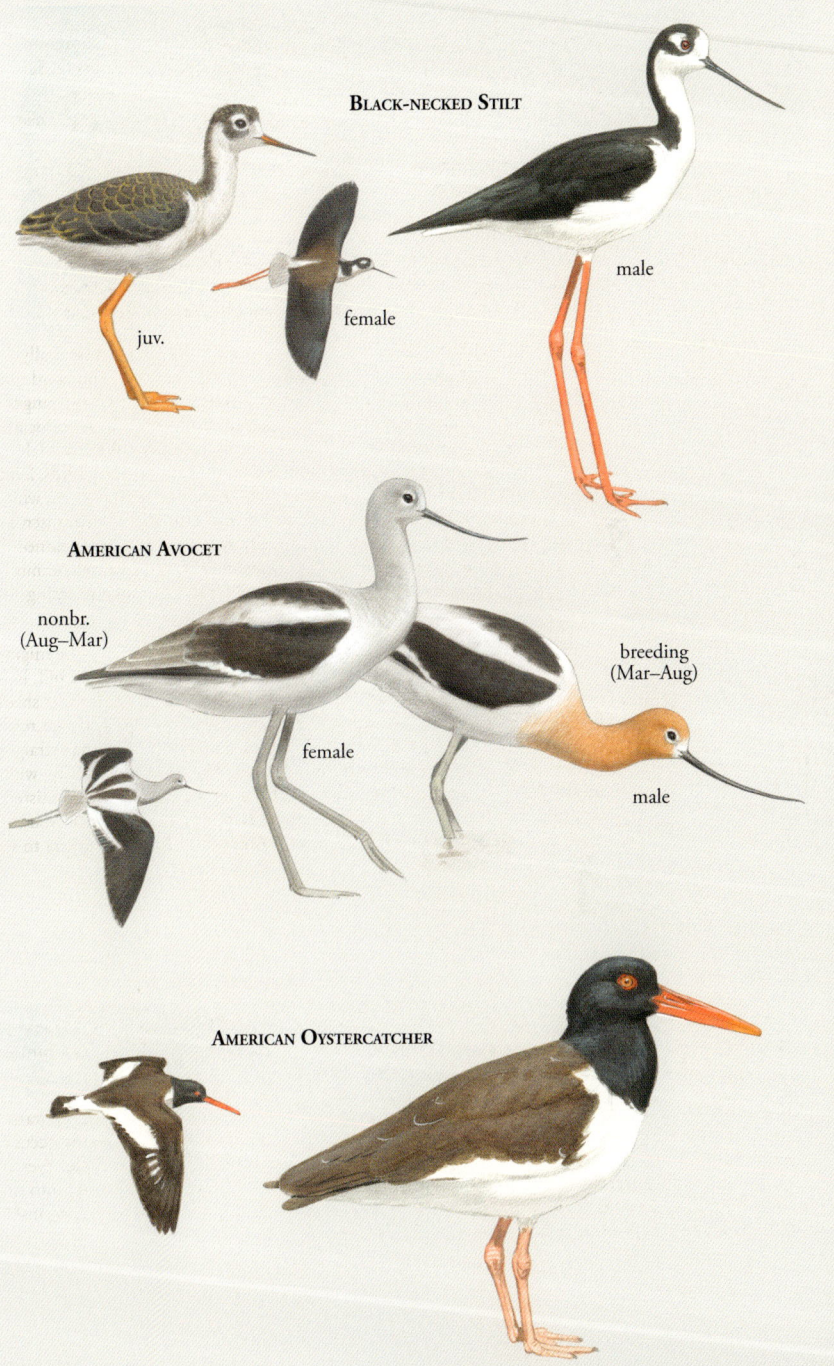

Black-necked Stilt

juv.

female

male

American Avocet

nonbr.
(Aug–Mar)

breeding
(Mar–Aug)

female

male

American Oystercatcher

PLOVERS (CHARADRIIDAE; 7+ SPECIES) Worldwide family of small to large shorebirds. Big eyes and short bills attest to visual hunting strategy, unlike probing and picking of sandpipers. Most migrants show seasonal variation; residents similar year-round. Ages usually differ, with juv. resembling nonbr.; attain adult appearance in 1st year. Sexes usually differ slightly, at least in breeding plumage, with males having more extensive black on face and underparts.

BLACK-BELLIED (GRAY) PLOVER *Pluvialis squatarola* 26.5–28cm. Large bulky plover of coastal habitats, from sandy beaches to mudflats, coastal lagoons. Locally in small flocks, associating readily with other shorebirds. Note large bulk, stout bill; in flight, note white rump, white wingstripe, black 'armpits.' Attains breeding plumage Feb–Apr, before migration. Juv. (Sep–Jan) resembles neater, browner version of nonbr., with streaked breast, finely spangled upperparts, cf. American Golden Plover. **SOUNDS:** Melancholy slurred whistles, *heéueeh* and *chweéee*; typically lower, more drawn-out than American Golden Plover. **STATUS:** Fairly common to common Aug–Apr on coast and cayes, smaller numbers May–Jul include oversummering nonbr. imms. Rare and sporadic inland, mainly during migration. (Breeds n. Eurasia and N America, winters almost worldwide.)

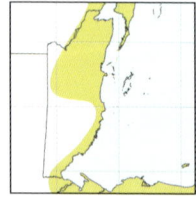

AMERICAN GOLDEN PLOVER *Pluvialis dominica* 24–25.5cm. Rare spring transient. Medium-size plover mostly found in open grassland, rice fields, at lakeshores, less often coastal habitats. Occurs as singles and small flocks, which may associate with other shorebirds. Smaller and slimmer than Black-bellied Plover, with slender bill, mostly dark upperparts (narrow whitish wingstripe), dusky underwings; also lacks small hind toe of Black-bellied. Juv. and nonbr. plumages rather dull above, not golden, with contrasting dark cap and whitish eyebrow. Attains breeding plumage Apr–May, at staging areas mainly n. of Belize. **SOUNDS:** Varied plaintive whistles. Fairly mellow *ch'weít* and flutier, more rolled *chweél*. **STATUS:** Uncommon to scarce and local Mar–May; unrecorded in fall, but possible. (Breeds N America, winters S America.)

*CAYENNE [SOUTHERN] LAPWING *Vanellus [chilensis] cayennensis* 34–36cm. Vagrant. Large spectacular shorebird, slowly spreading north from South America. No similar species in Belize: note wispy crest, black chest shield, big white wing panels in flight. Singles or rarely pairs could be found at wetlands, lakeshores, ranchland, and semi-open areas with ponds, often in areas with other waterbirds. Flies with unhurried bowed wingbeats suggesting a heron. **SOUNDS:** Can be noisy, including at night: varied strident barks and shrieking cries, mainly when disturbed and in display. **STATUS:** Vagrant, to date known only from Crooked Tree, where a single bird was present 2004–2007. Increasing in Central America and may become more frequent in Belize. (S America, spreading to Cen America.)

BLACK-BELLIED PLOVER

breeding
(Mar–Aug)

juv. (Sep–Jan)

nonbr.
(year-round)

AMERICAN GOLDEN PLOVER

nonbr.
(Mar–May)

juv. (Sep–Dec)

CAYENNE LAPWING

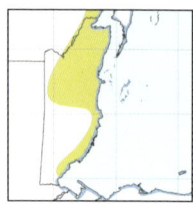

SEMIPALMATED PLOVER *Charadrius semipalmatus* 16.5–17.5cm. Nonbr. migrant to coastal habitats, especially muddy tidal flats and beaches; also occurs in mangroves, flooded fields, on lakeshores. Often in small flocks, which feed in well-spaced arrays but roost more tightly, often with other shorebirds. Note orange legs, wet-sand tone to upperparts, and stubby, orange-based bill. **SOUNDS:** Upslurred, slightly plaintive *ch'wièh* and sharper *ch'wiet!* Nasal bickering chatters in feeding interactions. **STATUS:** Fairly common Aug–May on coast and cayes, where rarely a few oversummer; rare to sporadically uncommon inland, mainly during migration. (Breeds N America, winters to S America.)

WILSON'S PLOVER *Charadrius wilsonia* 18–19cm. Small but big-billed plover, a local resident of sandy beaches, lagoon shores, mangrove flats; often breeds in areas with gravel. Mainly feeds by chasing down small crabs. Often in loose groups, associating with other shorebirds mainly at roosts. Note heavy black bill, pinkish legs. Male has black head and breast markings, female dark brown. **SOUNDS:** Sharp high *piik!* and clipped *pri-dik*; dry, buzzy chatters in interactions. **STATUS:** Uncommon to fairly common on coast and cayes, breeding locally in north, rarely in south; more widespread and numerous in fall–winter. (Americas.)

COLLARED PLOVER *Charadrius collaris* 14–15cm. Rare and poorly known in Belize. May be found in varied coastal habitats from beaches and river mouths to shrimp farms. Usually singles, rarely small groups, often apart from other shorebirds. Note neat, narrow black collar, rusty cheeks, fine dark bill, pinkish legs; lacks whitish hindneck collar of other small ringed plovers in Belize. Juv. plumage held briefly; no seasonal change in appearance, but patterns slightly veiled in fresh plumage. **SOUNDS:** Clipped sharp *pik*, suggesting Wilson's Plover; mellow rolled *krip* that can run into chatters. **STATUS:** Rare to very rare nonbr. visitor to coast and cayes, mainly Nov–Mar; rare sporadic breeder at coastal shrimp farms; very rare wanderer inland, Mar–Jun. (Mexico to S America.)

KILLDEER *Charadrius vociferus* 24–25.5cm. Relatively large, long-tailed, and often unabashedly noisy nonbr. migrant ringed plover, with distinctive double black breast band, bright rusty rump, and dark-tipped tail. Frequently away from water, in plowed fields, grasslands, other open habitats; rarely open beaches and tidal flats. Singles and small flocks may occur in any open grassy habitat, usually apart from other shorebirds. Ages/sexes similar. **SOUNDS:** Varied wailing and screaming cries, singly or in series, including a repeated, onomato-poeic *kill-deéu....* **STATUS:** Uncommon to fairly common Oct–Mar, a few birds sometimes from mid-Sep and into Apr. (Americas.)

SEMIPALMATED PLOVER

juv.
(Aug–Nov)

breeding
(Mar–Aug)

WILSON'S PLOVER

female

male

COLLARED PLOVER

juv.
(Jun–Sep)

adult

KILLDEER

SANDPIPERS (SCOLOPACIDAE; 24+ SPECIES) Nearly worldwide family of small to large shorebirds breeding mainly at high latitudes; no species breeds in Belize, but nonbr. imms. of several species may remain through the summer. Ages differ slightly to distinctly, with juv. usually resembling nonbr. plumage; attain adult appearance in 1st year. Sexes sometimes differ, at least in breeding plumage; females larger, longer-billed in most species.

*WHIMBREL *Numenius phaeopus* 35.5–43cm. Uncommon migrant. Large brown sandpiper with long decurved bill, found on beaches, at river mouths, coastal lagoons, estuaries, nearby fields; feeds by probing, at times while wading. Found singly or in small groups, readily associating with other shorebirds. Cf. larger and buffier Long-billed Curlew (cinnamon underwings, plainer face, and longer bill, but short-billed juv. approaches Whimbrel). Ages/sexes similar. SOUNDS: In flight, fairly rapid series of overslurred piping whistles, *pee-pee-pee...*, 6–9 notes/sec. Quavering fluty whistles in territorial interactions. STATUS: Uncommon Aug–Apr on coast and cayes; a few remain locally through summer. (Breeds n. N America, winters to S America.)

LONG-BILLED CURLEW *Numenius americanus* 45.5–58.5cm. Rare migrant. Very large, buffy-brown sandpiper with very long decurved bill (appreciably shorter on 1st-year). Favors tidal mudflats, also beaches, river mouths, coastal lagoons; feeds by probing, at times while wading. Usually singles, readily associating with other shorebirds; cf. Whimbrel, sleeping Marbled Godwit. Ages/sexes similar; 1st-year shorter-billed, adult female appreciably longer-billed than male. SOUNDS: Slightly shrieky hoarse *reeip* and slurred rising *hoooriep* in flight; plaintive quavering whistles and bubbling choruses. STATUS: Rare to very rare and irregular Aug–Apr on coast and cayes. (Breeds w. N America, winters to Cen America.)

MARBLED GODWIT *Limosa fedoa* 38–46cm. Rare migrant. Large buffy-brown sandpiper with long, slightly recurved, pinkish bill tipped dark. Favors tidal mudflats, beaches, river mouths, coastal lagoons; feeds by probing, often while wading in fairly deep water. Usually singles, readily associating with other shorebirds. Only godwit regularly seen in Belize; cf. Long-billed Curlew, especially when sleeping. Breeding plumage has variable dark barring on underparts. SOUNDS: Nasal, slightly crowing, *ah-ha* and *ahk*, at times in laughing series. STATUS: Rare and irregular Aug–Apr on coast and cayes, mainly in north. (Breeds N America, winters to Cen America.)

UPLAND SANDPIPER *Bartramia longicauda* 28–30.5cm. Scarce but widespread transient migrant in open grassy habitats, usually not around water. Found as singles or, especially in spring, in small groups, often apart from other shorebirds; feeds by picking in grass. Distinctive, medium-size, long-necked sandpiper with cryptic plumage, long tail, slender straight bill, yellowish legs. Little age/seasonal variation, but juv. neater and scaly above in fall. SOUNDS: Mainly in flight, a quick, slightly liquid whistled *whi-whit* and rippling *whi-whi-whuit*. STATUS: Uncommon and sporadic transient mid-Mar to mid-May, scarce late Jul to mid-Oct. (Breeds N America, winters S America.)

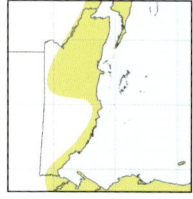

BUFF-BREASTED SANDPIPER *Calidris subruficollis* 18–20.5cm. Scarce transient migrant. Attractive small sandpiper of open habitats, from grassy fields and newly planted rice fields to lakeshores, rarely to sandy beaches; often not near water. Singles or rarely small flocks, at times associating loosely with other shorebirds, especially American Golden Plover, Upland Sandpiper. Walks with high-stepping gait, picking for food; can be confiding. Distinctive, with blank face, beady eye, plain buffy breast, yellow legs; white underwings contrast with buff body. Little age/seasonal variation but juv. notably fresh and scaly above in fall; male slightly larger than female. SOUNDS: Mostly silent; quiet low clucks on occasion. STATUS: Scarce and sporadic transient, mainly late Aug–Sep, late Mar to mid-May. (Breeds N America, winters S America.)

WHIMBREL

LONG-BILLED CURLEW

MARBLED GODWIT

breeding
(Mar–Aug)

nonbr.
(Aug–Mar)

juv.
(Aug–Oct)

UPLAND SANDPIPER

adult

juv. female
(Aug–Nov)

adult
male

BUFF-BREASTED SANDPIPER

GENUS *TRINGA* (5 species). Medium-size to large sandpipers with fairly long, often colorful legs and overall straight bills; upperparts typically have fine pale spotting or spangling in nonbr. plumages, vs. paler edging and scaly look typical of *Calidris* sandpipers. Often bob head or whole body when alarmed and tend to be fairly wary, flying off with whistled or yelping calls.

SOLITARY SANDPIPER *Tringa solitaria* 20.5–21.5cm. Medium-size migrant sandpiper of quiet freshwater ponds, lake edges, grassy wetlands; not on open mudflats or beaches. As the name suggests, does not associate strongly with other shorebirds, although small flocks can occur during migration. Often towers when flushed, wingbeats deep, quick, and swallow-like. Distinctive, with solitary habits, fairly long greenish legs, dark underwings in flight. Cf. Lesser Yellowlegs, Spotted Sandpiper. Little age/seasonal variation. **SOUNDS:** Slightly plaintive piping whistles in short series, *twee-weet* and *tweet-sweet-sweet*; brighter and flutier than Spotted Sandpiper, may suggest alarm call of Barn Swallow. **STATUS:** Uncommon to fairly common transient, late Jul–Oct, Mar–May, less numerous and more local in winter. (Breeds N America, winters Mexico to S America.)

LESSER YELLOWLEGS *Tringa flavipes* 24–25.5cm. Medium-size, long-legged sandpiper of varied fresh and saltwater habitats from small ponds and lakeshores to coastal lagoons, flooded rice fields; rarely open mudflats, beaches. Locally in flocks of 100s, often in same areas as Greater Yellowlegs. Feeds while wading, picking at water surface; rarely swims. Main confusion is with larger, stouter-billed Greater Yellowlegs, which is about willet-size, vs. Lesser, which is about dowitcher-size; Greater has louder, 'shouted' calls vs. mellower calls of Lesser. Also cf. Stilt Sandpiper, Solitary Sandpiper. **SOUNDS:** Downslurred whistled *tew* or *kyew*, often in short series suggesting Short-billed Dowitcher, rarely strident enough to suggest Greater Yellowlegs. **STATUS:** Fairly common to common Aug–Apr on mainland, scarce on cayes; more common and widespread in migration, Jul–Oct, Mar–May. (Breeds N America, winters to S America.)

GREATER YELLOWLEGS *Tringa melanoleuca* 29.5–31.5cm. Large, long-legged sandpiper of varied wetland habitats, much like Lesser Yellowlegs. Rarely in flocks of more than 20 birds. Feeds while wading; often dashes actively and sweeps bill side-to-side, vs. more sedate picking of Lesser Yellowlegs; rarely swims. Cf. willets, Lesser Yellowlegs. **SOUNDS:** Typical call a series of (usually 3–4) ringing downslurred whistle, *tchu-tchu-tchu*; repeated sharp yelp when agitated, *kyehw....* **STATUS:** Fairly common to common Aug–Apr on mainland, less numerous on cayes; more common and widespread in migration, Jul–Oct, Mar–May. A few may oversummer locally. (Breeds N America, winters to S America.)

*WESTERN WILLET *Tringa [semipalmata] inornata* 33–35.5cm. Large, rather stocky sandpiper of varied coastal habitats, shrimp farms near coast; feeds by picking and probing, often in shallow water. Willets as such are distinctive, but distinguishing Eastern from Western can be challenging (see below, under Eastern): note stout straight bill, whitish spectacles, gray legs; diagnostic wing pattern striking in flight. **SOUNDS:** Noisy. Varied, mainly 3- or 4-note mellow to loud shrieking whistles, *kri-wih-wih* and *krri-WI-WI-wihr*; nasal inflected *kyeh'eh* and short series, *kyeh-yeh-yeh*; alarm a sharp yapping *kyih!* at times repeated steadily. **STATUS:** Uncommon Aug–Apr on coast and cayes; more numerous and widespread in migration, Jul–Oct, Mar–May, when very rare inland; small numbers oversummer locally. (Breeds w. N America, winters to S America.)

*EASTERN WILLET *Tringa semipalmata* 32–34.5cm. Very similar to Western Willet, but slightly smaller and stockier overall, with deeper, blunter bill, shorter legs; breeding plumage averages darker and browner, with pinkish tinge to bill and legs. **SOUNDS:** Calls similar to Western, but average higher, less husky. **STATUS:** Scarce (overlooked?) coastal transient, Aug–Oct, Jan–Mar. (Breeds e. N America to Caribbean, winters S America.)

SOLITARY SANDPIPER

nonbr.
(Aug–Mar)

breeding
(Mar–Aug)

LESSER YELLOWLEGS

juv.
(Aug–Oct)

nonbr.
(Aug–Mar)

breeding
(Mar–Aug)

GREATER YELLOWLEGS

juv.
(Aug–Oct)

nonbr.
(Aug–Mar)

breeding
(Mar–Aug)

WESTERN WILLET

nonbr.
(year-round)

breeding
(Mar–Aug)

EASTERN WILLET

breeding
(Feb–Aug)

SPOTTED SANDPIPER *Actitis macularius* 16.5–18cm. Distinctive small sandpiper of varied fresh and saltwater habitats from lakeshores and rivers to estuaries and ponds, often with stony and rocky shores; perches readily on posts, mangrove branches. Usually single birds, rarely loose small groups in migration, typically apart from other shorebirds. Walks with almost constant bobbing of rear end; rarely wades in water. Flight typically low over water with stiff flicking beats of bowed wings. Note habits, white spur at chest sides; breeding plumage has variable black spotting below; juv. like nonbr. with narrow pale edgings to upperparts. **SOUNDS:** High, slightly plaintive to piping single notes and short phrases, *siit* and *swie-wie-wie…*, etc. **STATUS:** Fairly common but low density Aug–Apr; more widespread in migration, Jul–Sep, Mar–May. (Breeds N America, winters to S America.)

RUDDY TURNSTONE *Arenaria interpres* 22.5–24cm. Distinctive, chunky, medium-size sandpiper of coastal habitats, especially sandy beaches, jetties. Often in flocks, associating readily with other shorebirds. Pokes and overturns seaweed, stones, in search of prey. Note bright red-orange legs, dark breast patches, variegated upperparts with striking flight pattern. Breeding female has duskier head markings, duller upperparts than male; juv. (Aug–Oct) like nonbr. but with neat pale scaly edgings to upperparts, duller legs. **SOUNDS:** Sharp *kyew!* and relatively mellow *ch-tu*, can suggest Short-billed Dowitcher; varied bickering chatters. **STATUS:** Fairly common to common Sep–Apr in coastal areas, especially cayes; more widespread in migration, Jul–Oct, Mar–May, when very rare inland; small numbers oversummer locally. (Breeds n. Eurasia and N America, winters almost worldwide.)

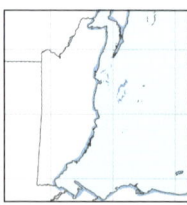

SANDERLING *Calidris alba* 18–19cm. A large 'small sandpiper' of sandy beaches, river mouths. In small groups, locally of 100s, less often singles; associates readily with other shorebirds, especially small sandpipers. Feeds by picking and probing, often along tideline. Note overall pale plumage with darker shoulder patch, medium-length straightish bill, black legs; lacks hind toe of other small and medium-size sandpipers. Breeding plumage rare in Belize. **SOUNDS:** Slightly nasal bright *kiip* and *whiik*. **STATUS:** Fairly common locally Sep–Apr on coast and cayes, more widespread during migration, late Jul–Oct, Apr–May. (Breeds n. Eurasia and N America, winters s. in New World to S America.)

WILSON'S PHALAROPE *Steganopus (Phalaropus) tricolor* 20.5–22cm. Scarce transient. Elegant, medium-size sandpiper of wetlands with open water, shrimp farms; rarely in coastal habitats. Feeds mainly while swimming, picking at water surface with fairly long fine bill; also feeds on shore, at times with tail cocked high, chasing flies. Mainly singles, rarely small flocks, associating readily with other shorebirds. Breeding plumage distinctive (male duller); nonbr. notably pale and silvery gray overall, in flight shows white rump. Cf. Lesser Yellowlegs and Stilt Sandpiper, which swim rarely. **SOUNDS:** Mostly silent, rarely uttering low muffled grunts. **STATUS:** Scarce to sporadically uncommon transient, Aug–Oct, Apr–May. (Breeds N America, winters mainly S America.)

WILSON'S SNIPE *Gallinago delicata* 25–26.5cm. Distinctive, medium-size, long-billed cryptic sandpiper of grassy marshes, lakeshores, other vegetated wetlands; not in open situations. Mostly seen feeding at marsh edges early and late in day or when flushed from grassy vegetation, usually as singles or loose aggregations. Flushed flight strong and erratic, often low initially then towering before dropping back to cover. All plumages similar, with striped face, bold buffy back stripes. **SOUNDS:** Usually utters low rasping *zzhek* when flushed. **STATUS:** Uncommon to locally fairly common Oct–Mar, a few from Sep and into Apr. (Breeds N America, winters to n. S America.)

SPOTTED SANDPIPER

breeding (Mar–Aug)

nonbr.
(Sep–Mar)

RUDDY TURNSTONE

nonbr.
(Aug–Mar)

male breeding
(Mar–Aug)

nonbr. (Aug–Apr)

SANDERLING

juv.
(Sep–Oct)

breeding
(Apr–Aug)

male

nonbr.
(Aug–Mar)

breeding
(Mar–Aug)

female

WILSON'S PHALAROPE

WILSON'S SNIPE

LEAST SANDPIPER *Calidris minutilla* 13–14cm. Commonest, most widespread small sandpiper in Belize, found in varied fresh and saltwater habitats from muddy ponds and lakeshores to tidal mudflats, rice fields; less often open beaches. Often in flocks, locally of 100s, mixing readily with other small sandpipers but often keeping to drier and more vegetated habitats than Western Sandpiper. Feeds by picking and probing; creeps along muddy shores with flexed legs rather than striding and wading in open water like Western Sandpiper. Note rather mouse-like demeanor, overall brownish plumage with white belly, medium-length decurved bill; yellowish legs often muddy. Western and Semipalmated Sandpipers larger and whiter-breasted, walk more upright, favor open habitats. SOUNDS: High, reedy trilled *krreep* and lower *krriit*; varied low trilling. STATUS: Fairly common to common Aug–Apr; more widespread in migration, mid-Jul to Sep, Mar–May. (Breeds N America, winters to n. S America.)

SEMIPALMATED SANDPIPER *Calidris pusilla* 14–15cm. Transient migrant. Small, rather compact sandpiper with short to medium-length bill, dark legs. Found in varied wetland habitats from coastal lagoons and mudflats to river mouths, flooded fields, lakeshores; associates readily with other small sandpipers. Feeds by probing and picking. From Western Sandpiper by structure, especially shorter, blunter-tipped bill (some female Semipalmated have longer bill, overlapping short-billed male Western); juv. more uniform above than juv. Western (vs. grayish with rusty scapulars). Cf. Least and other small sandpipers. SOUNDS: Fairly sharp *kyip*, lower *chrit*. STATUS: Fairly common to common transient, mid-Jul to Oct, late Mar–early Jun, most numerous in fall, especially Aug. Scarce and local in winter along mainland coast. (Breeds N America, winters Mexico to S America.)

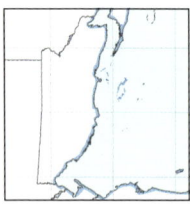

WESTERN SANDPIPER *Calidris mauri* 15–16.5cm. Small, rather long-necked sandpiper with relatively long, slightly decurved bill, black legs. Mainly in coastal habitats such as mudflats, river mouths, beaches, also shrimp farms; associates readily with other small sandpipers. Feeds by probing and picking, often wading up to its belly. Note contrasting gray and bright rusty tones on juv. and breeding plumages; cf. Semipalmated and Least Sandpipers. SOUNDS: High, scratchy, downslurred *chiit*, burry *chrrit*. STATUS: Fairly common to common Sep–Apr, especially on and near coast; more widespread in migration, late Jul–Oct, Mar–May, when rare inland; a few nonbr. imms. may remain through summer. (Breeds N America, winters to n. S America.)

WHITE-RUMPED SANDPIPER *Calidris fuscicollis* 17–18.5cm. Transient migrant. Small, very long-winged sandpiper of varied wetland habitats, from coastal mudflats to rice fields. Singles or small groups, seasonally 100s, mixing readily with other small sandpipers. Feeds by picking and probing, often wading in shallow water. Note long wings projecting past tail tip, lack of buffy plumage tones, voice; white uppertail coverts distinctive in flight. Cf. Semipalmated and Western Sandpipers. SOUNDS: Very high, slightly tinny, descending *jit*, easily missed. STATUS: Fairly common to common spring transient, mid-Apr to mid-Jun (mainly mid–late May); very rare and irregular in fall, mid-Sep to mid-Oct. (Breeds N America, winters S America.)

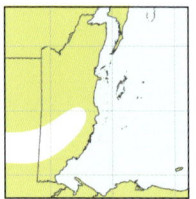

PECTORAL SANDPIPER *Calidris melanotos* 19.5–23cm. Transient migrant. Medium-size, cryptic sandpiper of marshy wetlands, flooded fields, lakeshores, coastal lagoons; rarely on open mudflats, beaches. Singles or small groups, at times associating with other shorebirds. Feeds by probing and picking, often near and within grassy vegetation. Note medium size (male appreciably larger than female), yellowish legs, clean-cut 'pectoral' demarcation between streaky brown breast and whitish belly. Little age/seasonal variation. SOUNDS: Rolled, slightly wet *krrip*, lower in male. STATUS: Fairly common to common late Jul–Oct, Mar–May. (Breeds N America, winters S America.)

LEAST SANDPIPER

juv.
(Aug–Oct)

nonbr. (Aug–Mar)

breeding
(Mar–Aug)

SEMIPALMATED SANDPIPER

juv.
(Aug–Oct)

nonbr.
(Aug–Mar)

breeding
(Mar–Aug)

WESTERN SANDPIPER

juv.
(Aug–Oct)

nonbr.
(Aug–Mar)

breeding
(Mar–Aug)

WHITE-RUMPED SANDPIPER

juv.
(Sep–Nov)

nonbr. (Aug–Mar)

breeding
(Mar–Aug)

PECTORAL SANDPIPER

juv. female
(Aug–Nov)

adult male

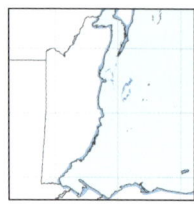

RED KNOT *Calidris canutus* 25.5–26.5cm. Medium-size, rather fat sandpiper of tidal mudflats, river mouths, coastal lagoons, beaches, shrimp farms. Singles, rarely small groups, mix readily with other sandpipers and Black-bellied Plovers. Feeds mainly by probing. Distinctive but rather nondescript in winter; body size slightly larger than a dowitcher, but bill and legs obviously shorter. Breeding plumage seen briefly in spring, less often on worn fall migrants. SOUNDS: Often silent. Nasal inflected *che'wet* and upslurred *wek* in flight. STATUS: Rare to uncommon and local Oct–Mar; more widespread during migration, late Aug–Oct, Mar–Apr. (Breeds n. Eurasia and N America, winters s. in New World to S America.)

STILT SANDPIPER *Calidris himantopus* 20–21.5cm. Medium-size, rather long-legged sandpiper of fresh and brackish marshes, shrimp farms, coastal lagoons; rarely estuaries, beaches. Often in groups, feeding and roosting with Long-billed Dowitchers; less often singles mixed among other shorebirds. Feeds by probing while wading up to its belly, rear end typically raised steeply out of water. Slightly smaller and paler gray than Long-billed Dowitcher, with whitish brow, shorter black bill with slightly drooped tip; in flight note white rump, feet projecting well past tail tip. Also cf. Lesser Yellowlegs. SOUNDS: Mostly silent; quiet gruff grunts and clucks on occasion. STATUS: Uncommon to fairly common locally Sep–Apr; more widespread in migration, Aug–Oct, Mar–May. (Breeds N America, winters Mexico to S America.)

SHORT-BILLED DOWITCHER *Limnodromus griseus* 24–28cm. The default coastal dowitcher, favoring tidal flats, river mouths, mangroves; also freshwater areas in migration. Often in flocks, associating with other shorebirds; sometimes with Long-billed Dowitcher, mainly in migration. Distinguished with care from Long-billed, which favors fresh water, is darker overall in all plumages, with broader dark tail barring; juv. Long-billed has narrow chestnut edging to upperparts, lacks buff notching on tertials and coarse mottling on scapulars of juv. Short-billed; nonbr. Long-billed has darker chest without dusky spotting of Short-billed, heavier dark flank barring; breeding Long-billed solidly rusty below, barred on sides of breast, with white tips to scapulars. Also cf. Stilt Sandpiper. Populations of Short-billed differ in breeding plumage, some populations extensively rusty below. SOUNDS: Mellow *ch-tu* or *ch-tu-tu* and longer variations; quality recalls Lesser Yellowlegs. STATUS: Fairly common Aug–Apr on coast and cayes; more widespread in migration, mainly Jul–Sep, Apr–May, when may occur rarely inland. (Breeds N America, winters to S America.)

LONG-BILLED DOWITCHER *Limnodromus scolopaceus* 25.5–29cm. The default inland dowitcher, where may occur locally in 100s at freshwater wetlands, flooded rice fields; also shrimp farms, mangroves, sewage ponds, but rarely open tidal flats. Habits much like Short-billed Dowitcher (see that species for ID). Fresh spring breeding plumage shown opposite; becomes appreciably darker by fall. SOUNDS: High sharp *kiik!* recalls yipping call of Black-necked Stilt; usually given singly by birds in flight or in rapid short series by flushed birds; sometimes lower-pitched and trebeled, *kih-tii-tii,* inviting confusion with Short-billed Dowitcher. STATUS: Fairly common locally Sep–Apr; more widespread in migration, late Aug–Oct, Apr–May. (Breeds N America, winters to Cen America.)

RED KNOT

juv.
(Sep–Oct)

breeding
(Mar–Aug)

nonbr.
(Aug–Mar)

STILT SANDPIPER

breeding
(Mar–Aug)

juv.
(Aug–Oct)

nonbr.
(Aug–Mar)

SHORT-BILLED DOWITCHER

nonbr.
(Aug–Mar)

variation

breeding
(Mar–Aug)

juv.
(Aug–Oct)

LONG-BILLED DOWITCHER

nonbr.
(Aug–Mar)

juv.
(Sep–Nov)

breeding
(Mar–Aug)

HERONS (ARDEIDAE; 17 SPECIES) Worldwide family of typically long-necked, long-legged birds with dagger-like bills; usually near water. Fly with neck retracted in bulge, unlike ibises, spoonbills, storks. Ages differ or similar, attain adult appearance in 1–3 years. Bare-parts often brighten or change color strikingly for brief periods at height of breeding season. Most species nest colonially in trees and marshes, often in mixed-species aggregations. Usually quiet except when disturbed or interacting at colonies; bitterns and tiger heron, however, have 'songs.'

WESTERN CATTLE EGRET *Bubulcus ibis* 45–53cm. Fairly small, compact white heron usually found in fields and farmland near cattle, horses, tractors, which flush up prey; also flooded fields, lakeshores, but not habitually out in wetlands, unlike most herons and egrets. Social, usually in groups, locally to 100s. Distinctive, with stocky shape, yellow bill, dark legs; cinnamon-buff plumes on crown, chest, and back most extensive on breeding adult, can be absent on 1st-year; at height of breeding, bill and lores flush salmon, legs scarlet. Fledgling bill can be blackish, soon like adult. In flight, note rather stocky shape without pronounced neck bulge of Snowy Egret. SOUNDS: Gruff clucks and grunts, mainly when nesting. STATUS: Fairly common to common mid-Sep to early May; smaller numbers occur locally through the summer, but breeding in Belize apparently unconfirmed. (Americas, Africa, and W Eurasia.)

SNOWY EGRET *Egretta thula* 49–59cm. Elegant, fairly small white egret of varied wetland habitats from beaches and river mouths to small ponds, mangroves. Singles or small groups, rarely low 100s, hunt while wading or waiting. Often dashes actively in shallows. Note slender black bill and contrasting yellow lores; adult has shaggy crest, ornate back plumes, yellow feet contrasting with blackish legs; lores flush orange at height of breeding. Fledgling bill can be yellowish, tipped black, soon like adult; 1st-year has greenish-yellow feet and hind edge to legs, cf. 1st-year Little Blue Heron. SOUNDS: Varied guttural rasps and croaks; higher, more nasal than American Great Egret. STATUS: Fairly common to locally/seasonally common Sep–Apr; smaller numbers occur locally through the summer, but breeding in Belize apparently unconfirmed. (Americas.)

***AMERICAN GREAT EGRET** *Ardea [alba] egretta* 84–99cm. The only large white heron in most of Belize, found in a wide variety of wetland habitats from roadside ditches and flooded fields to coastal lagoons, mangroves. Mostly singles or small groups, but 100s can gather at favored feeding lagoons with aggregations of other wading birds. Note wholly white plumage, yellow bill, blackish legs and feet; adult has long ornate back plumes, variable black on maxilla in breeding season, bright green lores at height of breeding. Cf. scarce Great White Heron in n. coastal areas and cayes. SOUNDS: Varied, deep guttural calls, often with creaky quality; average less stentorian than Great Blue Heron; lower, harsher than Snowy Egret. STATUS: Fairly common to locally/seasonally common Sep–Apr; smaller numbers occur through the summer and breeds locally in coastal areas. (Americas.)

***GREAT WHITE [GREAT BLUE] HERON** *Ardea [herodias] occidentalis* 102–127cm. Coastal. Very large white heron of cayes, mangroves, beaches. Usually singles, sometimes in association with other wading birds. Only likely confusion is with slightly smaller, less heavily built American Great Egret: note Great White Heron's stout bill with variable dark on maxilla, mostly gray lores, pale tibia; adult has thin wispy crest. At height of breeding, adult lores flush blue, bill is wholly orange-yellow, legs pink. Presumed hybrids with Great Blue Heron resemble pale Great Blue with extensively white face. SOUNDS: Much like Great Blue Heron. STATUS: Scarce and sporadic on n. cayes, very rare visitor on mainland coast s. to Dangriga. (Caribbean region.)

1st-year/
nonbr.

fledgling

**WESTERN
CATTLE EGRET**

breeding

**imm. Little
Blue Heron
(p. 66)**

adult

SNOWY EGRET

1st-year

**AMERICAN
GREAT EGRET**

white
**Reddish Egret
(p. 66)**

1st-year

breeding

1st-year

GREAT WHITE HERON

adult

**GREAT WHITE ×
GREAT BLUE HERON**

LITTLE BLUE HERON *Egretta caerulea* 51–61cm. Fairly small dark heron of varied wetland habitats, from river mouths and mangroves to flooded fields, shrimp farms; more often found at isolated small ponds and ditches than other egrets. Singles or small groups, locally 100s at favored feeding areas, often mixed with other wading birds. Hunts mainly by waiting, slow stalking. Adult smaller and darker than Reddish Egret, with blue-gray base to bill, yellow-green legs; at height of breeding, bill base and lores flush blue, legs and feet blackish. 1st-year Little Blue easily passed off as Snowy Egret but slightly stockier, with bluish-gray face and black-tipped bill, yellowish-green legs and feet, fine dark tips to outer primaries. Piebald molting birds seen frequently for a few months, mainly May–Aug. **SOUNDS:** Much like Snowy Egret, but often slightly raspier. **STATUS:** Fairly common to locally/seasonally common, mainly Aug–Apr; small numbers oversummer locally, mainly imms., but not known to breed in Belize. (Americas.)

TRICOLORED HERON *Egretta tricolor* 56–66cm. Handsome, 'snake-necked' heron of varied wetland habitats, from river mouths and mangroves to flooded fields, coastal lagoons, shrimp farms. Singles or small groups, rarely more than 20 or so birds together; often with other wading birds. Hunts by waiting, often coiled and hunched low in shallow water, and by active dashing. Distinctive, but perhaps better named 'bicolored' heron, with overall dark slaty-gray plumage and contrasting white belly; also note long slender bill, yellowish face; at height of breeding, bill base and lores flush violet-blue, legs and feet pinkish red. 1st-year has rusty head and neck sides, like adult in 2nd year. **SOUNDS:** Relatively high, drawn-out, squawking *aaáah*, at times in short series; also lower nasal calls. **STATUS:** Fairly common, mainly Oct–Mar, especially in coastal areas, with smaller numbers through summer; breeds locally in north. (Americas.)

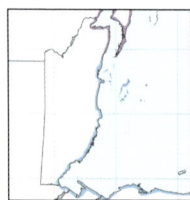

REDDISH EGRET *Egretta rufescens* 66–77cm. Rather large, dimorphic egret (white morph uncommon) of coastal habitats from lagoons and river mouths to beaches, mangroves. Singles or small aggregations, often with other wading birds; hunts by dashing actively, often raising wings, also by stalking and waiting. Note fairly large size, slender bill (pink base most of year on adult), dark legs; at height of breeding, lores and legs flush cobalt blue, bill base brighter pink. Dark morph adult has shaggy rusty head and neck plumes; white morph wholly white (all ages). 1st-year dark morph slaty gray overall with variable rusty tinge to neck, dark bill, staring whitish eyes; cf. adult Little Blue Heron. **SOUNDS:** Low moaning groans and grunts; mostly quiet away from colonies. **STATUS:** Uncommon to fairly common on n. coast and cayes, breeding locally; rare visitor to s. coast, very rare wanderer inland. (N America to Caribbean, winters to S America.)

GREAT BLUE HERON *Ardea herodias* 102–127cm. The only very large dark heron in Belize, widespread in fresh and saltwater habitats, from beaches and mangroves, to lakeshores, flooded fields, roadside ponds and ditches. Singles or local concentrations up to 50 or so birds, often in areas with other herons, egrets, storks, ibises. Hunts by waiting and slow stalking. Flight heavy, with slow deep wingbeats, neck retracted in a bulge like other herons; neck can be extended briefly after takeoff. Adult has clean white crown, shaggy neck and back plumes, clean plumage; at height of breeding, lores flush blue-gray, bill bright orange-yellow. 1st-year has black crown, duskier plumage with pale edgings to upperparts; 2nd-year like dull adult, crown partially to mostly white. **SOUNDS:** Flight call a loud, explosive *rrek!* Other varied sounds include deep throaty croaks, often with raspy, complaining quality. **STATUS:** Fairly common but mostly low density, mainly late Sep–Apr, with smaller numbers through summer; breeds locally in north. (N America to Mexico, winters to S America.)

Little Blue Heron

juv.

breeding

adults

molting
1st-years

Tricolored Heron

breeding

adults

juv.

dark morphs

white
morph

adults

1st-year

Reddish Egret

breeding

1st-year

Great Blue Heron

adult

BLACK-CROWNED NIGHT HERON *Nycticorax nycticorax* 56–64cm. Chunky, mainly nocturnal heron often seen during the day in varied fresh and brackish wetland habitats. Hunts along edges and in shallow water, waiting and stalking slowly for fish. Roosts mainly in trees, often not deeply hidden, at times with or near Yellow-crowned Night Herons. Rather compact in flight, with short foot projection past tail tip, cf. longer-legged, more lightly built Yellow-crowned. Adult plumage distinctive; juv./1st-year from Yellow-crowned by shape, yellow-based pointed bill, shorter legs, coarser pale spots and streaks on upperparts. Cf. Boat-billed Heron. 2nd-year like duller, browner version of adult. At height of breeding, adult lores become black, legs flush bright salmon-red. SOUNDS: Flight call a gruff barking *worhk!* or *wahk*, with rising inflection. Other low guttural calls when nesting and disturbed. STATUS: Uncommon to fairly common, mainly Oct–May, but scarce on cayes; small numbers occur locally in summer, but breeding as yet unconfirmed in Belize. (Worldwide except Australasia.)

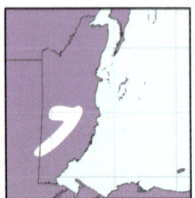

YELLOW-CROWNED NIGHT HERON *Nyctanassa violacea* 51–59cm. Mainly nocturnal but often seen during the day, in varied brackish and freshwater habitats, especially mangroves. Main food is crabs, and regularly hunts away from water, such as in coconut plantations and on open beaches. Roosts mainly in trees, often not deeply hidden. More lightly built than Black-crowned Night Heron with stouter blackish bill, longer neck, longer legs; in flight, feet project well past tail tip. Adult distinctive. Cf. juv./1st-year Black-crowned (nestling Yellow-crowned has yellow at bill base, soon darkens). 2nd-year like duller, browner version of adult. At height of breeding, adult lores become black, legs flush bright salmon-red. SOUNDS: Flight call a slightly grating *owhr* or *kyowh*, higher and more nasal than Black-crowned Night Heron, typically with more downward inflection. Low guttural clucks when nesting and disturbed. STATUS: Fairly common to common in coastal areas and on cayes, less numerous and more local inland; breeds locally, mainly in coastal belt and on cayes. (Americas.)

*BOAT-BILLED HERON *Cochlearius cochlearius* 46–53cm. Distinctive nocturnal heron of fresh and brackish marshes, coastal lagoons, slow-moving rivers, mangroves; spends the day roosting, well hidden in trees; tends to leave roost later than night herons. Hunts at night along shorelines and in shallow water. Flight direct with slightly stiff wingbeats emphasizing the upstroke, subtly distinct from steadier bowed wingbeats of night herons. Note very broad bill with distensible pouch, big dark eyes. Slightly smaller and more compact than Black-crowned Night Heron, with shorter toe projection in flight; adult has contrasting black underwing coverts. Juv. lacks pale spots and streaks of juv. night herons. SOUNDS: Usually silent in flight at dusk. At roost when disturbed, and when nesting, utters varied clucks and chatters, often with chuckling cadence, such as *kuh-kuh kuk-kuh ku-kah*. STATUS: Uncommon to fairly common locally, especially in coastal belt lowlands. (Mexico to S America.)

BLACK-CROWNED NIGHT HERON

juv.

adults

YELLOW-CROWNED NIGHT HERON

juv.

adults

BOAT-BILLED HERON

juv.

adults

LEAST BITTERN *Ixobrychus exilis* 28–31cm. Very small, retiring heron of fresh and brackish marshes with tall reeds, rushes, lakes with bordering reedbeds; migrants may occur in any wet habitat, including mangroves. Rarely wanders far from cover; mainly hunts from perch over water, clambering easily through reeds. Daytime flights usually short and low over reedbeds, with fairly quick wingbeats; flushes from close range, legs often dangling, and usually flies a short distance before dropping back to cover. Distinctive (note big buff wing panel in flight); cf. larger, stockier, and overall dark juv. Green Heron. Male cap and back black, female dark brown. At height of breeding, male lores flush scarlet, legs bright orange. **SOUNDS:** Common call a raspy barking *kyeh-kyeh-kyeh-kyeh-kyeh*, slowing slightly at end, given irregularly, sometimes when flushed; suggests Clapper Rail but shorter, not pulsating. In breeding season, 'song' a fairly rapid series of about 5–10 muffled coos, *cuh-cuh-cuh-cuh-cuh*, often repeated steadily and can be given at night; may suggest Black-billed Cuckoo song but lower, huskier. **STATUS:** Scarce to uncommon Sep–Apr; smaller numbers occur more locally in summer, and breeds locally. (Americas.)

GREEN HERON *Butorides virescens* 38–43cm. Small dark heron of varied wetland habitats from mangroves and cayes to small roadside ponds, extensive wetlands, shrimp farms, often with wooded edges and cover nearby. Usually singles, locally a few birds concentrated at good feeding sites, and often apart from other herons. Hunts by waiting or slow stalking, often hunched motionless on low branch or other perch over water. Distinctive, given small size, overall dark plumage; note rusty neck sides, bright yellow legs. Juv. has streaked neck, much like adult by end of 1st year. At height of breeding, adult lores become black, legs flush bright salmon-orange. **SOUNDS:** Flight call a clipped, slightly explosive yapping *kyah!* or *kyowh*, often in short series when flushed; hollow, low clucking series when agitated, *kuh-kuh-….* In breeding season, 'song' is a low, frog-like growl, *reeohr*, repeated. **STATUS:** Fairly common to common, but low density, Sep–Apr; smaller numbers occur through summer and breeds locally, including cayes. (N America to Panama.)

AGAMI HERON *Agamia agami* 66–76cm. Scarce. Very long-billed tropical heron of shady forested wetlands, mangroves, quiet forest streams and ponds. Usually solitary, quiet, and stealthy; often rather shy and usually apart from other herons, but breeds locally in colonies. Easily overlooked, dark plumage blending well with shady habitats. Note very long slender bill, rather short dull legs, habits. Adult stunning, with silvery blue-gray filigree neck plumes, chestnut neck and belly, deep oily-green upperparts; attains plush silvery-gray crest in breeding season, when lores and throat flush bright red. Imm. distinctive, with very long bill, brown face, neck and upperparts, pale belly. **SOUNDS:** Mostly quiet. Territorial call a low, throaty, purring growl, about 1–2 secs, repeated every few secs. **STATUS:** Generally scarce and local, but seasonally fairly common around breeding colonies and in late winter at Crooked Tree. (Mexico to S America.)

Least Bittern

juv.

female

male

Green Heron

juv.

adults

variation

Agami Heron

breeding

1st-year

adult

BARE-THROATED TIGER HERON *Tigrisoma mexicanum* 71–82cm. Large, heavyset, very long-necked but rather short-legged heron of marshy wetlands, wooded swamps, coastal lagoons, occasionally mangroves. Usually close to wooded cover or tall reeds, and perches readily in trees (unlike bitterns). Usually seen as singles, locally loose concentrations up to 10 or so birds, sometimes in fairly open situations along with other herons, storks, ibises; hunts by waiting and slow stalking. Flight heavy, with stiff wingbeats emphasizing upstroke; note broad wings, very short foot projection. Usually identified readily by large size, dark plumage, short legs; also note naked yellow throat. Cf. Great Blue Heron, Pinnated Bittern. Adult has gray face, black cap, fine barring on neck and upperparts; juv./1st-year has coarsely barred and banded ('tiger-striped') plumage; 2nd-year like duller, more coarsely barred adult. **SOUNDS:** Flushed birds often give a low, guttural grunting *koh koh....* In breeding season, especially at dusk or during night, 'song' is a far-carrying, steadily repeated, throaty grunt or roar, at times with paired cadence from dueting or countersinging birds, *rrohr, rrohr...*, 10 notes/10–15 secs. **STATUS:** Fairly common to uncommon. (Mexico to nw. Colombia.)

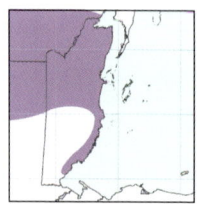

PINNATED BITTERN *Botaurus pinnatus* 64–76cm. Large but cryptic, buffy-brown heron of fresh marshes with tall grassy vegetation, reedbeds, flooded rice fields. Mostly skulking, but sometimes hunts in fairly short and open roadside grass and flooded fields. Tends to crouch or freeze when alarmed, but at times flushes from close range; flight heavy, with wingbeats mainly above body plane, more reminiscent of a tiger heron than of American Bittern. Pale buffy plumage with coarsely marked upperparts and boldly barred neck sides distinctive; white tufts at chest sides usually concealed but flared in display. **SOUNDS:** Flushed birds give gruff *owhk*, singly or in short series. 'Song' a very low-pitched, deep booming or gulping *uungh*, often preceded by a few liquid clucks and at times repeated steadily; quality recalls American Bittern. **STATUS:** Scarce to uncommon and seemingly rather local. (Mexico to S America.)

AMERICAN BITTERN *Botaurus lentiginosus* 51–61cm. Rare winter migrant. Skulking, cryptic brown heron of fresh and brackish wetlands, especially marshes with taller reeds and rushes, lakes bordered by reedbeds. May be seen feeding at edges, but rarely in open; when alarmed, crouches in shorter vegetation or stretches neck up in taller reeds to blend in. Note relatively plain brown neck sides and upperparts, boldly striped foreneck with blackish whisker bordering throat. In flight, note large, strongly curled feet projecting past tail tip. Darker, plainer, and colder brown overall than larger Pinnated Bittern. Cf. 1st-year night herons, which are bigger headed, spotted pale above. **SOUNDS:** Flushed birds give slightly guttural *woh*, less emphatic than night herons. Flight call a throaty *arhk*, lower, softer than night herons. **STATUS:** Rare and sporadic Oct–Mar; might be found anywhere in suitable habitat. (Breeds N America, winters to Cen America.)

Bare-throated Tiger Heron

1st-year

2nd-year

1st-year

adult

Pinnated Bittern

American Bittern

IBISES AND SPOONBILLS (THRESKIORNITHIDAE; 4+ SPECIES)
Worldwide family of elegant wading birds; ibises have slender decurved bills, spoonbills (p. 76) have flattened bills with spatulate tips. Ages differ, sexes similar; attain adult appearance in 2–3 years. Build stick nests in trees; often nest colonially with other waterbirds.

WHITE IBIS *Eudocimus albus* 53–63cm. Distinctive wading bird of mangroves, tidal flats, coastal lagoons, shrimp farms, less often freshwater marshes; main food is crabs, other crustaceans. Gregarious, usually in groups, but rarely more than 100 birds away from roosts; feeds by probing and picking. Flies with neck and legs outstretched, wingbeats fairly quick and shallow, interspersed with brief glides; often flies in lines and Vs, usually not mixed with other waterbirds. Distinctive, with decurved reddish-pink bill, pinkish-red legs; small black wing-tips of adult often hidden at rest. Juv. slightly bulkier than Glossy Ibis, told by white underparts, bright pink bill; attains white upperparts gradually over 1st year. Cf. much larger Wood Stork. **SOUNDS:** Mostly quiet; low grunts when flushed and in interactions. **STATUS:** Fairly common on cayes and along coast, breeding locally; scarce to sporadically fairly common visitor inland, especially at shrimp farms. (Americas.)

GLOSSY IBIS *Plegadis falcinellus* 53–63cm. Slender, wholly dark ibis of varied fresh and brackish wetland habitats, from roadside ponds to extensive marshes, flooded fields, rarely mangroves. Locally in flocks, also singles, associating readily with other wading birds; feeds by probing. Flies with neck outstretched and slightly drooped, legs trailing (feet project noticeably farther on male), wingbeats fairly stiff and shallow, often interspersed with brief glides on bowed wings. Distinctive, but cf. imm. White Ibis, Limpkin. Also cf. distant flying Neotropic Cormorant, which can suggest a dark ibis. **SOUNDS:** Mostly silent; low nasal grunts when flushed. **STATUS:** Uncommon to rare, mainly Sep–May in north, but records year-round and might occur anywhere; not known to breed in Belize. (Old World, N America, Caribbean region.)

Very similar **White-faced Ibis** *Plegadis chihi* (51–61cm) first confirmed 2010s in Belize (may be rare but regular, overlooked). Best told from Glossy Ibis by red eyes, pinkish facial skin with narrow paler borders; breeding adult attains feathered white border to face; juv. and young imm. perhaps not safely separable from Glossy. Birds not seen well are best logged as *Plegadis* sp. = species unidentified. Hybrid White-faced × Glossy Ibis known from N America and could occur in Belize, shows intermediate face patterns.

LIMPKINS (ARAMIDAE; 1 SPECIES) New World family related to cranes and rails.
Ages differ slightly, sexes similar; like adult in 1st year.

*LIMPKIN *Aramus guarauna* 58–64cm. Distinctive large wading bird of fresh marshes, lakes, swampy woodland, wet savanna. Usually near cover and can be retiring, at other times singles or locally groups feed out in the open, probing for snails; often perches in trees. Flight distinctive but usually not prolonged, with neck outstretched, stiff snappy wingbeats with emphasis on upstroke, like a crane. Nothing very similar, but cf. smaller and more lightly built Glossy Ibis. Note Limpkin's stout straight bill with black tip, variably white-spattered plumage, habits. Juv. has neater and smaller white markings than adult. **SOUNDS:** Varied loud trumpeting, screaming, and clucking calls, especially early and late in the day and at night, including a rolled *krrrreeah* and honking *krrrrowh*; clipped barking or *owhk!* in alarm and sharp, piercing *bihk, bihk, …* when agitated. **STATUS:** Fairly common to locally common, especially in coastal belt lagoons. (Mexico and Caribbean region to S America.)

WHITE IBIS

juv.

adult

imm.

GLOSSY IBIS

imm./nonbr.

breeding

imm./nonbr.

breeding

WHITE-FACED IBIS

LIMPKIN

ROSEATE SPOONBILL *Platalea ajaja* 71–79cm. Stunning pink wading bird of coastal marshes and lagoons, mangroves, less often fresh wetlands. Associates readily with other wading birds, especially White Ibis. Feeds by filtering out food with its bill tip. Flies with neck and legs outstretched, wingbeats fairly quick and shallow, interspersed with brief glides; regularly soars, at times in kettles with vultures, Wood Storks. Distinctive bill shape and pink plumage should preclude confusion. Juv. much paler, with fine dusky wing-tips, feathered head that changes to naked pale greenish over 1st year; 2nd-year like duller version of adult. **SOUNDS:** Mostly quiet except when nesting; colonies produce low clucking and chuckling calls. **STATUS:** Uncommon to fairly common in coastal belt lowlands, breeding locally in north; uncommon to scarce and sporadic visitor inland. (Americas.)

STORKS (CICONIIDAE; 2 SPECIES) Mainly Old World family of large to very large wading birds with stout pointed bills, long broad wings often used for soaring. Ages differ, sexes similar; adult appearance attained in 2–3 years. Build bulky stick nests in trees; Wood Stork usually colonial. Utter low grunts and hisses, also bill-rattling noises; mostly silent away from nest.

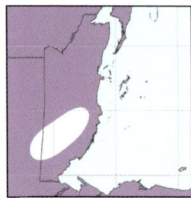

WOOD STORK *Mycteria americana* 89–101cm. Large, stout-billed wading bird of fresh and brackish wetlands, from coastal lagoons, flooded fields, and roadside ponds (especially when drying up) to mangroves, shrimp farms, wooded swamps. Feeds by wading and probing, often in association with other waterbirds; perches readily in trees. Often soars on mid–late morning thermals, at times high overhead with vultures; flocks tend to wheel in somewhat disorganized kettles, not strongly synchronized, cf. distant American White Pelicans. Juv./1st-year has creamy bill, downy head feathering; head becomes naked and attains adult pattern over 1–2 years. **STATUS:** Fairly common but local and nomadic; breeds locally, mainly in north, with wandering birds possible anywhere, including flying high over forested regions. (Americas.)

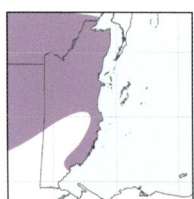

JABIRU *Jabiru mycteria* 130–153cm. Huge, the largest flying bird in the Americas, and difficult to misidentify. Singles or locally small groups in open wetland habitats, savannas, flooded rice fields, often in areas with aggregations of egrets, herons, other wading birds, when large size usually obvious. Flight strong, with smooth wingbeats, neck outstretched; soars occasionally. Juv./1st-year has upperparts edged silvery gray-brown, but can looks whitish overall at a distance; red on neck duller than adult, often some downy whitish feathering on head. 2nd-year like dull adult, with scattered brownish feathers on upperparts. **STATUS:** Uncommon to locally/seasonally numerous in north, also south locally in coastal belt. Breeds mainly late Nov–Apr; more widespread with post-breeding dispersal, May–Oct, when rare and sporadic in south and has reached Ambergris Caye. (Mexico to S America.)

ROSEATE SPOONBILL

imms.

adults

adults

adults

JABIRU

WOOD STORK

juv.

78

RAILS, GALLINULES, ALLIES (RALLIDAE; 13 SPECIES) Worldwide family of very small to medium-size marsh birds. Can be divided into more skulking rails and crakes, and more conspicuous gallinules and coots, which are often seen swimming. Ages differ, sexes usually similar; precocial downy young of all species are black and fuzzy; attain adult appearance in 1st year. Flight can be surprisingly strong, but migrate at night and rarely seen in flight unless flushed.

SORA *Porzana carolina* 20.5–21.5cm. Fairly small migrant crake of varied marshy habitats, from roadside ditches and extensive wetlands to mangroves, lakeshores. More conspicuous than other crakes, out at marsh edges any time of day, but dashes quickly for cover when alarmed; tail often cocked in a point, near vertical. No similar species in Belize: note short yellowish bill, black face, buffy-white wedge under tail. Imm. attains adult appearance over 1st winter. **SOUNDS:** Slurred, slightly nasal squealing *kee-ur,* bright clipped *keek,* and short rolled squeal run into a descending, slow-paced whinny, *kreeh, dee-de-du-du…,* 1.5–4 secs. **STATUS:** Fairly common to locally common, Sep–Apr. (Breeds N America, winters to S America.)

RUDDY CRAKE *Laterallus ruber* 14.5–15.5cm. The common small crake of Belize, heard often but seen rarely. Inhabits varied grassy and marshy habitats from damp fields and roadside ditches to extensive wetlands with taller reeds. Sometimes comes out at edges, mainly early and late in day, when can be confiding; rarely flies. Distinctive, with slaty-gray hood, bright ruddy plumage, no paler barring or spotting. **SOUNDS:** Rising and falling purring trills, typically preceded by soft, hesitant piping notes audible at close range, *whiit, whitt… urrrrrrrr…,* mainly 1.5–6 secs, prolonged series up to 30 secs. Song a hard, clicking *tk* or *chk,* often doubled, and alternated with short rasps, *tk, tk-tk ehrr, tk-tk, tk ehrr….* **STATUS:** Fairly common to common, generally less numerous in the north. (Mexico to Costa Rica.)

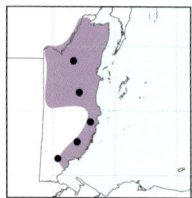

*NORTHERN BLACK RAIL *Laterallus jamaicensis* 13–14cm. Small elusive crake of bunch grass in pine savanna, seasonally flooded grassy areas, marshes. Skulking and difficult to see; may approach within a few feet but be hidden like a mouse, even in short vegetation. Note dark bill, whitish spotting on upperparts, pinkish legs; cf. Gray-breasted Crake. **SOUNDS:** Sings mainly at night. Song 2–3 bright piping notes followed by a short growl, *hii'kii-derr* or *hii'kii-kii-durr,* every 1–6 secs. Gruff rasping churring series when agitated, *jhehr-jhehr….* **STATUS:** Scarce and local presumed breeding resident. (Americas.)

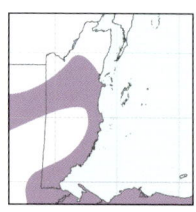

GRAY-BREASTED CRAKE *Laterallus exilis* 14.5–15.5cm. Small elusive crake of varied grassy and marshy habitats, from extensive wetlands with taller reeds to overgrown weedy fields, roadside ditches; often in same areas as commoner Ruddy Crake. Skulking and difficult to see; may approach within a few feet but be hidden like a mouse, even in short vegetation. Note lime-green base to bill, barred flanks, yellowish legs; cf. Northern Black Rail. **SOUNDS:** Song a fairly rapid series of (usually 2–9) high piping notes, often with a soft introductory note audible at close range: *tik, dee-dee…,* 3–4 secs; easily passed off as a frog. Churring rattles recall Ruddy Crake but lower and drier, less bubbling, often shorter. **STATUS:** Uncommon to fairly common locally. (Mexico to S America.)

YELLOW-BREASTED CRAKE *Laterallus (Hapalocrex) flaviventer* 12.5–13.5cm. Very small crake of fresh marshes, especially with emergent and floating vegetation, from small ponds to large wetlands. Typically retiring but may feed out at edges, walking on floating vegetation. Flushes silently from underfoot, legs dangling, and flies a short distance back to cover. Distinctive, with dark cap and eyestripe, buff face and breast, very long yellow to pinkish toes. **SOUNDS:** Song a plaintive, slightly metallic, 2-note whistled phrase, *chieh-dii,* every 1.5–5 secs, including at night; less often a single *chieh.* Quiet short rattle, *chrrrt,* when agitated may run into series of low, gruff, rasping scold notes, *zzheh-zzheh….* **STATUS:** Scarce to uncommon and local, mainly in north, but might occur anywhere in suitable habitat. (Mexico to S America.)

SORA

imm.
(Aug–Oct)

RUDDY CRAKE

juv.

juv.

NORTHERN BLACK RAIL

GRAY-BREASTED CRAKE

YELLOW-BREASTED
CRAKE

SPOTTED RAIL *Pardirallus maculatus* 25.5–28cm. Medium-size, strikingly patterned rail of fresh marshes, rice fields, overgrown wet ditches. Often rather skulking and difficult to see, emerging mainly at dusk along marsh edges, but at times feeds unconcerned in open situations at any time of day. Relatively large size, striking plumage, and bright bare parts distinctive. Juv. may be dimorphic: some like duller version of adult, others dark overall, but soon attaining distinctive barring and spotting. SOUNDS: Overslurred, rough screeching *rreéah*, about 0.5 sec. Varied series of nasal pumping grunts or rough screeching clucks, often disyllabic with overall descending cadence, 2–5 notes/ sec. Agitated birds give very low, hollow, stuttering or purring chatter suggesting a distant outboard motor, 12–15 notes/sec. STATUS: Scarce to sporadically fairly common. (Mexico to S America.)

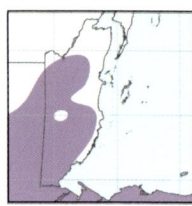

UNIFORM CRAKE *Amaurolimnas concolor* 20.5–22.5cm. Medium-size, rather plump-bodied rail of damp grassy thickets, *Heliconia* stands, rainforest understory. Typically skulking and rarely seen, walks and runs with fairly upright stance suggesting a tinamou; usually detected by voice. Note short yellowish bill, overall dull rusty plumage with no barring, bright pinkish-red legs. SOUNDS: 'Song' mainly early and late in day, and at night, a series of (usually 7–20 or more) easily imitated, upslurred whistles, each 0.5–1 sec, series at times intensifying then fading quickly: *tuuth tuuth….TOO'IH TOO'IH…. too-ih*; at a distance, only loudest sections audible. Sharp clucking *plik!* when disturbed. STATUS: Scarce and local. (Mexico to S America.)

CLAPPER RAIL *Rallus longirostris* 30.5–38cm. Large rail of mangroves. Mostly seen as singles at edges early and late in the day; prefers to run rather than fly, but can flush explosively from underfoot. No similar species in Belize. Note blue-gray cheeks, dull pinkish-cinnamon breast, overall grayish upperparts with dark streaks; legs typically dull pinkish but can fade to drab yellowish, flush to bright orange-red. Juv. has dingy grayish neck and underparts with little or no flank barring, duller bill and legs; like adult by fall. SOUNDS: Loud clucking chatters year-round, typically with intensifying and fading rhythm *kah-kah…*, about 6–8 notes/sec, can be prolonged in duet; quality suggests Least Bittern chatter but faster-paced, more prolonged, with pulsating rhythm. Male advertising call an emphatic gruff *kehk!* repeated at varying rates, steadily at about 2–4 notes/sec. Alarm call a throaty growl, *rehrr*, can be preceded by a cluck. STATUS: Fairly common locally on cayes and along mainland coast. (Mexico and N America to Caribbean.)

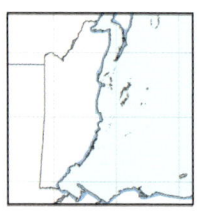

RUFOUS-NECKED WOOD RAIL *Aramides axillaris* 29.5–31.5cm. Handsome large rail of mangroves. Mostly skulking, but may be found feeding quietly on open mudflats near cover. Note bright rusty head and breast, blue-gray upper back, black undertail coverts. Juv. has dirty grayish head and neck, imm. like dull adult into mid-winter. SOUNDS: Sharp clucks, at times in steady series, *kek-kek…*, 4–5 notes/sec. STATUS: Uncommon to locally fairly common late Aug–May on inshore cayes and along mainland coast; not known to breed in Belize. (Mexico to S America.)

RUSSET-NAPED [GRAY-NECKED] WOOD RAIL *Aramides [cajanea] albiventris* 38–43cm. Very large, handsome rail of fresh and brackish wetland habitats, typically with wooded or other cover nearby; locally in mangroves. Distinctive, but in coastal mangroves cf. Rufous-necked Wood Rail. Singles or pairs walk along shorelines and marsh edges, often boldly in rather open situations, including roadsides; tends to be elusive when calling. Runs when disturbed, rarely flies. Juv. much like adult, but with grayish belly, duller bare parts. SOUNDS: 'Song' mainly early and late in day in spring–summer, a rollicking, clucking and shrieking phrase repeated over and over, *koh-koh KWIdi-KWIdi…*, or *koh koh-koh ki-KWIdi-KWIdi…*, at times in crazed-sounding duets; cf. Spotted Wood Quail. Sharp clucking shrieks when disturbed. STATUS: Fairly common in lowlands, uncommon to scarce at higher elevations. (Mexico to Costa Rica.)

SPOTTED RAIL

UNIFORM CRAKE

CLAPPER RAIL

RUFOUS-NECKED WOOD-RAIL

juv. (Aug–Oct)

RUSSET-NAPED WOOD-RAIL

AMERICAN COOT *Fulica americana* 35.5–40.5cm. Distinctive, chunky, rather duck-like waterbird of varied fresh and brackish habitats, from lakes and coastal lagoons to wetlands with open water; very rarely marine waters. Locally in flocks of 100s or low 1000s; elsewhere singles and small groups, associating readily with other waterbirds. Feeds while swimming by diving and upending, and while walking on shore, when big lobed toes distinctive. Juv. (not known in Belize) has extensively whitish head and neck, attaining adult appearance over 1st winter. **SOUNDS:** Varied nasal clucks, rough quacks, and gruff chatters, mainly in interactions; typically lower and gruffer than calls of Common Gallinule; most commonly a gruff *krreh*, and nasal, slightly trumpeting *puh!* **STATUS:** Uncommon to locally common, mainly mid-Oct to Apr; rare and sporadic in summer, mainly Crooked Tree. (Breeds N America to Cen America, winters to n. S America.)

COMMON GALLINULE (MOORHEN) *Gallinula galeata* 33–35.5cm. Freshwater marshes, small ponds, and lakes with reeds and other vegetation cover, flooded fields, drainage ditches. Singles or small groups, usually near cover. Feeds mainly while walking, also swims readily. Distinctive, with white stripe along sides, big white undertail-covert wedges; attains adult appearance over 1st year. **SOUNDS:** Varied. Sharp, overslurred nasal *kek!* and sharp clucking *kuh*, at times repeated steadily or in short series; accelerating then slowing series of sharp clucks with laughing or cackling cadence, tailing off with disyllabic notes, 2–5 secs, *k'keh-kehkehkeh keh keh, k'eh, k'eh....* **STATUS:** Uncommon to locally fairly common, mainly Nov–Mar, with smaller numbers from Sep and into Apr; rare and sporadic in summer, and may breed locally; migrants occasional on cayes. (Americas.)

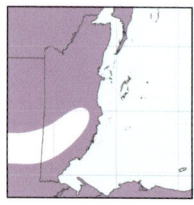

PURPLE GALLINULE *Porphyrio martinicus* 30.5–33cm. Spectacular large 'rail' of vegetated freshwater habitats, especially with floating vegetation, overhanging bushes, reedbeds. Feeds in open areas, especially on floating vegetation, but often rather shy; runs or flies to cover when disturbed and climbs readily in marshy vegetation, often fairly high in bushes. Distinctive, with long, bright yellow legs and toes, large single white wedge on undertail coverts; note bluegreen sheen on juv. wings; attains adult appearance over 1st year. Cf. Common Gallinule. **SOUNDS:** Varied sharp clucks, typically more nasal, lower than Common Gallinule and often repeated in faster series, sometimes speeding and slowing slightly; overslurred wailing or nasal trumpeting *meéah*, about 0.5 sec, often doubled or in short series, can suggest Limpkin. **STATUS:** Uncommon to fairly common but local on mainland; scarce migrant on cayes. (Americas.)

JACANAS (JACANIDAE; 1 SPECIES) Small pantropical family of rather rail-like shorebirds with very long toes that help them walk on floating vegetation; thus also known as lily-trotters. Ages differ, sexes similar but female larger; male takes care of young. Attain adult appearance in about 1 year.

NORTHERN JACANA *Jacana spinosa* 21.5–24cm. Distinctive bird of varied wetland habitats, especially with floating vegetation, from lakes and roadside ponds to flooded fields, rarely mangroves. Found as singles or locally loose aggregations, rarely to 100 or more birds. Walks with high-stepping gait, flies with quick stiff wingbeats and short glides, long legs and toes trailing; bright yellow wings striking in flight. No similar species in Belize. Female averages larger than male, with bigger shield, longer spurs at bend of wing. **SOUNDS:** Loud screechy clucks and chatters, often in rapid series; shrill raucous chatter when flushed. **STATUS:** Fairly common to locally common. (Mexico to Panama.)

AMERICAN COOT

imm.

COMMON GALLINULE

imm.

PURPLE GALLINULE

imm.

NORTHERN JACANA

1st-year

NEW WORLD QUAIL (ODONTOPHORIDAE; 3 SPECIES) Handsome
'chickens' of brushy, scrubby, and forested habitats; mainly terrestrial, but roost in trees and bushes. Often in groups in nonbr. season; can be elusive, but unwary when acclimated. Ages differ; adult appearance attained within a month or so by near-complete molt. Sexes differ or similar. Often detected by voice.

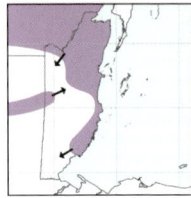

YUCATAN (BLACK-THROATED) BOBWHITE *Colinus nigrogularis* 19–21cm. Small quail of open grassy and scrubby habitats, overgrown clearings, farmland, mainly in the pine savanna zone; other quail occur in humid forest. In pairs or groups, tending to run for cover or flush noisily when disturbed. No similar species in habitat. **SOUNDS:** 'Song' (mainly spring–summer) a slightly hoarse to ringing whistled *hu-WHIIH!* ('bob-WHITE!') or *h-hoo-WHUIIH!* every 4–15 secs. Ringing, hollow whistled *whíuh*, year-round, at times in steady series or rhythmic couplets; varied nasal clucks and twitters. Low wing whirr when flushed. **STATUS:** Fairly common to uncommon; spreading locally with deforestation. (Mexico to Nicaragua.)

SINGING QUAIL *Dactylortyx thoracicus* 20–23cm. Small cryptic quail of semi-deciduous forest, not in open habitats; heard more often than seen. Usually in small groups, at times scratching loudly in leaf litter, which draws attention; can be confiding. Runs or freezes in preference to flying. Distinctive if seen well; cf. larger Spotted Wood Quail. **SOUNDS:** 'Song' an arresting series of (usually 3–15) clear to tremulous whistles, starting hesitantly and accelerating into rapidly repeated (usually 2–8×), short, loud, rollicking choruses: *wheeeu, wheeeur wheeeur-…kí-kí-i-weér, kí-kí-i-weér…*; introductory whistles recall Little Tinamou, may be given singly throughout day. Liquid twitters when agitated. **STATUS:** Uncommon and seemingly local; reported but unconfirmed from Maya Mts. (Mexico to Honduras.)

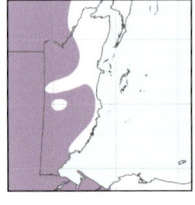

SPOTTED WOOD QUAIL *Odontophorus guttatus* 23–26cm. Chunky dark quail of humid forest; heard more often than seen. In pairs or small groups on shady forest floor; usually runs or freezes when alarmed; will flush at close approach. Distinctive if seen well: note blackish throat, white droplet spots on underparts; often raises and lowers crest while walking around (female lacks orange base to crest). Underparts vary from brown to dull rusty. Cf. Singing Quail. **SOUNDS:** 'Song' a prolonged rhythmic chorus, can last a min or longer, mainly early and late in day; typically alternated pairs of 3–4-syllable whistled phrases, such as *hu-WA-hoo hu-widl, hu-WA-hoo hu-widl.…* Single birds give rhythmic, steadily repeated *hu'wik!* or *hoh-woh*, 1–2 phrases/sec, can lead into song choruses. **STATUS:** Fairly common to uncommon. (Mexico to Panama.)

CHACHALACA, GUAN, CURASSOW (CRACIDAE; 3 SPECIES) Neotropical
family of large, long-tailed, frequently arboreal gamebirds known as cracids. Ages usually similar, sexes similar or (in Great Curassow) strikingly different.

PLAIN CHACHALACA *Ortalis vetula* 41–56cm. Frequently conspicuous and noisy inhabitant of brushy woodland, forest edge, gallery forest, second growth, overgrown clearings; but can be elusive when quiet. Usually in groups, feeding from ground to high in trees; typically 'sings' from mid–upper levels in trees. Often seen flying across roads and clearings with sweeping glides, neck outstretched, tail fanned on landing, when size and pale tail tips might suggest Brown Jay, but no truly similar species in Belize. Plumage averages paler below and on tail tips in drier n. areas. **SOUNDS:** Noisy, mainly in early morning. Loud raucous calls shrieky to burry, typically a 3-syllable *kuh-kuh-ruh* or *cha-cha-lac*, characteristically repeated in chanting, rhythmic duets. Gruff purring and growling when agitated; other shrieking and honking chatters can suggest *Amazona* parrots. **STATUS:** Fairly common to common. (Mexico and s. Texas to nw. Costa Rica.)

female

male

YUCATAN BOBWHITE

female

male

SINGING QUAIL

brown
female

males

SPOTTED WOOD QUAIL

PLAIN CHACHALACA

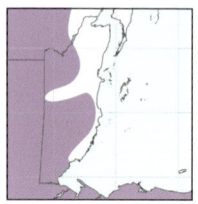

CRESTED GUAN *Penelope purpurascens* 81–91cm. Hunted, and rarely seen near habitation away from protected areas. Very large, dark, arboreal cracid of humid forest. In pairs or small groups, usually in canopy and at fruiting trees and shrubs; rarely on ground. Distinctive, much larger and darker than Plain Chachalaca, with big red throat wattle, gray face and bill, white streaks on neck and breast, erectile bushy crest. Often looks blackish overall in poor light. **SOUNDS:** Far-carrying honking and yelping cries, often repeated tirelessly early and late in day, such as *yoink yoink…*, 2–3/sec, at a distance might suggest a pygmy owl; can break into more excited screaming when disturbed. Loud rushing or crashing of wings when flushed from trees. In breeding season around dawn, muffled wing drumming rattle produced in display flight through canopy: about 5 accelerating notes, a brief pause, then about 12 fast-paced notes, *drruh drruh-drruh-drruhdrruh, drruhdrruh…*, 2.5 secs total. 'Song' in early morning an overslurred rasping honk or nasal whistle that fades into a moan, *reówhn* or *wheéohrr*, every 2–8 secs; at a distance might suggest a frog. **STATUS:** Fairly common (in remote areas and where not hunted) to scarce and widely extirpated. (Mexico to S America.)

GREAT CURASSOW *Crax rubra* 76–92cm. Hunted, and rarely seen near habitation away from protected areas. Distinctive, very large, mostly terrestrial cracid of humid forest. Singles, pairs, or small groups feed on ground, less often in fruiting trees; males may sing from perch in canopy. Wary in most areas and runs off quickly when disturbed, but can become acclimated when not hunted; rarely flies. Note very large size, curly crest, big yellow bill knob of male. Female plumage variable, with less common dark and barred morphs present in most populations. Juv. resembles adult of respective sex, even when only about 60% of adult size; male develops yellow bill knob over a few months. **SOUNDS:** 'Song' at any time of day, a very low-pitched, almost subliminal booming *uhmmm*, and variations; alarm call a sharp, overslurred piping *wheep!* **STATUS:** Widely extirpated or scarce in many areas; fairly common to common locally in protected and remote areas. (Mexico to S America.)

TURKEY AND ALLIES (PHASIANIDAE; 1 SPECIES) Largely Old World family of gamebirds, mainly in temperate habitats. Ages/sexes differ mainly in size, like adult in 1st year.

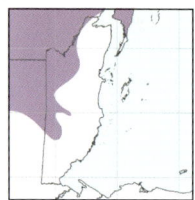

OCELLATED TURKEY *Meleagris ocellata* Male 92–102cm, female/imm. 66–84cm. Hunted, and rarely seen near habitation away from protected areas. Very large, spectacular gamebird of forest and edge, adjacent brushy clearings, overgrown corn fields; not domesticated. Singles or family groups, usually wary and not near people, but can be acclimated where not hunted; typically runs when alarmed, rather than flying; roosts in trees. No similar species in range, but beware of domesticated Wild Turkeys, none of which approach Ocellated in plumage. Male larger and brighter than female, with erectile forehead wattle, long leg spurs. **SOUNDS:** In display, male gives slightly nasal pumping calls that accelerate into a 'gobble,' sounding more like a motor scooter starting up than the gobble of Wild Turkey. Female gives low clucks. **STATUS:** Uncommon to scarce where heavily hunted, still fairly common to common in some protected and remote areas. (Mexico to n. Guatemala and n. Belize.)

CRESTED GUAN

barred morph

GREAT CURASSOW

females

dark morph

male

rusty morph

OCELLATED TURKEY

males

in display

TINAMOUS (TINAMIDAE; 4 SPECIES) Neotropical family of fat-bodied, seemingly tail-less, terrestrial birds, heard far more often than seen. Ages similar, sexes often differ slightly, with females brighter. Feed and nest on ground (eggs notably lustrous), but roost in trees and shrubs.

THICKET TINAMOU *Crypturellus cinnamomeus* 25.5–29cm. Medium-size, brightly marked tinamou of semi-deciduous forest, adjacent second growth. Walks quietly and often quickly on ground, but may be detected when rustling in dry leaves; prefers to run or freeze when alarmed; flushes explosively with whirring wings. Local overlap with similar-size Slaty-breasted Tinamou, which has grayish head and breast, male lacks barring above, female barring less contrasting than on Thicket. Sings year-round, but much less often in cooler winter months. **SOUNDS:** Song a melancholy, drawn-out pure whistle, 0.5–1 sec; subtle modulations produce 2–3-parted cadence, *whoo-oo* and *whoo-oo-oo*; often 15 secs or longer between songs. **STATUS:** Fairly common to common. (Mexico to Costa Rica.)

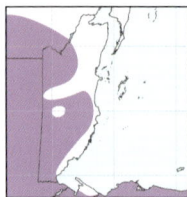

SLATY-BREASTED TINAMOU *Crypturellus boucardi* 25.5–28cm. Medium-size, rather dark tinamou of humid forest. Walks quietly and often quickly on shady forest floor; prefers to run or freeze when alarmed. Note medium-size, grayish breast, red legs; cf. Thicket Tinamou, mainly in drier habitats. Also cf. smaller Little Tinamou, much larger Great Tinamou. **SOUNDS:** Far-carrying, low, mournful, slightly tremulous, drawn-out *whoooo-oo-oooo*, about 2 secs; usually at least 15 secs between songs; rarely a shorter *hoo-oo*; quality recalls sound made by blowing across the top of a bottle. **STATUS:** Uncommon to fairly common. (Mexico to Costa Rica.)

LITTLE TINAMOU *Crypturellus soui* 20–24cm. Small tinamou of humid second-growth thickets, grassy forest edge, rarely forest interior. More retiring and usually harder to see than larger tinamous, aided by its denser habitat. Note unbarred plumage, greenish-yellow legs, small size; cf. Slaty-breasted Tinamou, Uniform Crake (p. 80); also cf. voice of Great Tinamou, Singing Quail. **SOUNDS:** Drawn-out clear whistle, 1–2 secs, often rises slightly and falls off with slight quaver, *wheee-eeeerr*, every 3–15 secs, at times in paired or tripled cadence (pairs and trios? counter-singing?). Less frequently a series of (usually 4–8) clear, tremulous whistles, about 1/sec; series intensifies then ends abruptly, recalls introduction to Singing Quail song: *wheeeéh, wheeeer, wheeeer….* **STATUS:** Fairly common in lowlands, scarce at higher elevations. (Mexico to S America.)

GREAT TINAMOU *Tinamus major* 38–46cm. Very large, heavy-bodied tinamou of humid forest, especially shady and fairly open rainforest interior. Runs or freezes when alarmed, and flushes explosively with whirring wings. Note large size, rather plain plumage (some individuals browner overall than shown) with narrow whitish eyering, grayish legs. Cf. song of Little Tinamou. **SOUNDS:** Haunting song mainly early and late in day, also at night: intensifying series of (usually 2–6) pairs of mournful, quavering whistles with variable introduction of 1–4 shorter notes, *whi, who, who who-hooorr, who-hooorr….* Lower, more mournful than Little Tinamou, with different cadence. Flushed birds make rapid whistling twitter as they fly off. **STATUS:** Fairly common (where not hunted) to uncommon or scarce (where hunted, and at higher elevations). (Mexico to S America.)

female

male

THICKET TINAMOU

female

male

SLATY-BREASTED TINAMOU

LITTLE TINAMOU

male

female

GREAT TINAMOU

NEW WORLD VULTURES (CATHARTIDAE; 4 SPECIES) Small New World family of carrion feeders. Differ from hawks and eagles in weaker bills and feet, lighter build, naked and often colorful heads. Ages differ slightly to distinctly, sexes similar; adult appearance attained in 1 year for smaller species, 3–4 years for King Vulture.

*NORTHERN TURKEY VULTURE *Cathartes aura* 66–76cm, WS 165–183cm. Widespread and familiar. Occurs in a wide variety of habitats, mainly in open, semi-open, and wooded country; tends to avoid heavy forest. Roosts communally in towns on large pylons and towers. Soars and glides for long periods with little flapping, often tilting side to side; wings held in a shallow V and wingbeats deep, elastic. Distinctive in most of range, but cf. Lesser Yellow-headed Vulture of savanna and marshes. Juv. attains red head within a few months. **SOUNDS:** Occasional soft clucks and hisses. **STATUS:** Common to fairly common throughout. (N America to Cen America, winters to S America.)

LESSER YELLOW-HEADED VULTURE *Cathartes burrovianus* 56–62cm, WS 148–165cm. Northern Turkey Vulture look-alike of savanna, marshes, coastal lagoons; mostly seen quartering low over marshes, other open country. Shape and flight manner much like Northern Turkey Vulture, but tail slightly shorter, less graduated, and rarely flies high overhead. Slightly smaller and blacker than Northern Turkey Vulture, with more-contrasting whitish shafts on outer primaries. Adult head colors gaudy and striking when seen up close. Juv. has mostly grayish head with ghosting of adult pattern, attains adult colors within a few months. Usually silent. **STATUS:** Uncommon to fairly common locally in north (including Ambergris Caye) and s. along coastal strip. (Mexico to S America.)

BLACK VULTURE *Coragyps atratus* 56–66cm, WS 140–158cm. Widespread, familiar, and distinctive. Occurs in towns, villages, other open and semi-open country, often near water and along rivers, at garbage dumps; tends to avoid heavy forest. Often roosts on cliffs, pylons. Typically soars higher than Northern Turkey Vulture, often in disorganized kettles; flies with wings flatter than Northern Turkey Vulture, and wingbeats very different: stiff and hurried, usually in short bursts. Commonly in groups on the ground, where hops and ambles readily, unlike Northern Turkey Vulture, which is awkward on the ground. Juv. has smoother head and darker bill than adult, like adult in 1–2 years. **SOUNDS:** Occasional sneezy hisses. **STATUS:** Common to fairly common. (N America to S America.)

KING VULTURE *Sarcoramphus papa* 71–82cm, WS 176–193cm. Large spectacular vulture of rainforest and adjacent areas, locally in pine savanna. Mostly seen as singles or pairs in flight, often high overhead; sometimes in kettles with other vultures. Groups of 10–20 birds may gather locally at carcasses, when dominates smaller Black Vulture. Adult distinctive (cf. flying Wood Stork and Northern White Hawk at long range). Juv. lacks white primary panels of Black Vulture but has whitish mottling on underwing coverts, broader and longer wings held flatter with tips curled up. Attains adult plumage in about 3–4 years. Usually silent. **STATUS:** Uncommon to locally fairly common, mainly in more remote and forested areas, not near towns. (Mexico to S America.)

NORTHERN TURKEY VULTURE

juv.

adults

LESSER YELLOW-HEADED VULTURE

juv.

adults

BLACK VULTURE

adult

juv.

KING VULTURE

juv.

adults

DIURNAL RAPTORS (42+ SPECIES) Popular, widespread assemblage of predatory birds, comprising 3 families: Ospreys (Pandionidae), Hawks (Accipitridae; including kites, eagles), Falcons (Falconidae; including caracaras). Range from dainty American Kestrel to huge Harpy Eagle. Ages differ, sexes different or similar (females larger, strikingly so in some species); adult appearance attained in 1–2 years for falcons and medium-size to large hawks, up to 4–5 years for big eagles. ID of some raptors straightforward but others can be challenging, compounded by age variation and highly variable plumages. Appreciation of behavior, habitat, and structure (wing shape on flying birds, relative wing and tail lengths on perched birds, head and bill size) often more helpful than colors and patterns, although those should always be noted.

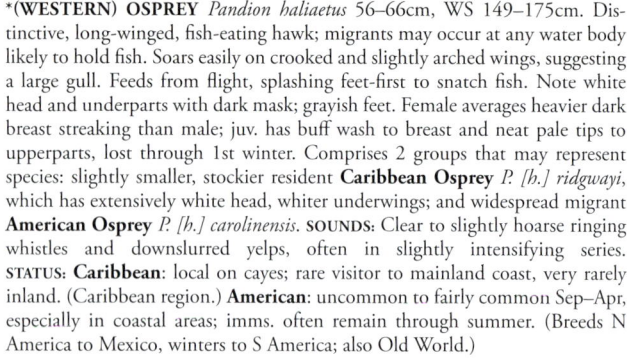

*(WESTERN) OSPREY** *Pandion haliaetus* 56–66cm, WS 149–175cm. Distinctive, long-winged, fish-eating hawk; migrants may occur at any water body likely to hold fish. Soars easily on crooked and slightly arched wings, suggesting a large gull. Feeds from flight, splashing feet-first to snatch fish. Note white head and underparts with dark mask; grayish feet. Female averages heavier dark breast streaking than male; juv. has buff wash to breast and neat pale tips to upperparts, lost through 1st winter. Comprises 2 groups that may represent species: slightly smaller, stockier resident **Caribbean Osprey** *P. [h.] ridgwayi*, which has extensively white head, whiter underwings; and widespread migrant **American Osprey** *P. [h.] carolinensis*. SOUNDS: Clear to slightly hoarse ringing whistles and downslurred yelps, often in slightly intensifying series. STATUS: **Caribbean**: local on cayes; rare visitor to mainland coast, very rarely inland. (Caribbean region.) **American**: uncommon to fairly common Sep–Apr, especially in coastal areas; imms. often remain through summer. (Breeds N America to Mexico, winters to S America; also Old World.)

LAUGHING FALCON *Herpetotheres cachinnans* 46–56cm, WS 79–94cm. Distinctive, large-headed and long-tailed snake-eating raptor of forest edge, savanna, open country with scattered taller trees and forest patches. Often perches on prominent snags, fence posts, utility poles. Does not soar; flight direct, with hurried stiff wingbeats that show off bold buff wing panels. Note creamy-buff head and underparts with broad 'bandit mask,' boldly banded tail. Ages/sexes similar. SOUNDS: Far-carrying crowing and laughing calls often reveal its presence; calls past sunset, unlike Collared Forest Falcon, which has similar calls. Steady *hah hah…* or *haáh, haáh…*, about 1 note/sec, or a slightly slower *wáko, wáko,…*; series can be prolonged, sometimes breaking in maniacal laughing cackles. STATUS: Uncommon to fairly common. (Mexico to S America.)

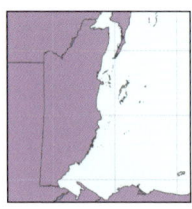

CRESTED CARACARA *Caracara plancus* 48–59cm, WS 115–132cm. Striking scavenger of open country with scattered trees, savanna. Associates readily with vultures; walks confidently on ground. Flies with steady strong wingbeats, soars occasionally. Note black cap, brightly colored face, long yellow legs. Flight pattern distinctive, with big white wing panels, white tail tipped black. Juv. browner than adult, with streaked (not barred) breast, duller face; 2nd-year like adult, duller, messier version of adult. SOUNDS: Throaty clucks and creaky rattles given mainly in interactions. STATUS: Rare and local, but increasing; first documented Belize in 2000, spreading with deforestation from Guatemala and Mexico. (US to S America.)

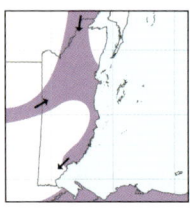

NORTHERN HARRIER *Circus hudsonius* 46–56cm, WS 99–117cm. Distinctive but scarce winter migrant hawk of open country, including shrimp farms, rice fields. Note white rump and broad dark trailing edge of secondaries in all plumages. Often seen in flight, quartering and sailing low over ground; perches on ground, fence posts, rarely high up. Flies with buoyant and smooth wingbeats and glides easily, wings held in a shallow V. 1st-year most often seen in Belize, has mostly unstreaked cinnamon underparts, often fading to pale buff by spring. Usually silent in Belize. STATUS: Uncommon to rare Oct–Mar, mainly in the north; occasionally lingers into Apr. (Breeds N America, winters to n. S America.)

OSPREY

adults

**Migrant
(American)**

**Resident
(Caribbean)**

**LAUGHING
FALCON**

CRESTED CARACARA

adults

adults

1st-year

NORTHERN HARRIER

1st-year

females

male

MIGRANT ACCIPITERS

Mostly favor forest habitats and seen infrequently except when migrating. Relatively short wings and long tails good for maneuvering among trees, where mostly hunt birds. Bicolored Hawk (a resident accipiter of tropical forest) is on p. 112.

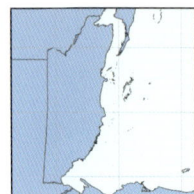

SHARP-SHINNED HAWK *Accipiter striatus* 28–36cm, WS 51–64cm. Rare winter migrant to wooded and forested habitats; rarely perches in open situations. Note small size, quick snappy wingbeats, relatively short rounded wings, long tail with overall squared tail tip, thin legs. Main confusion is with larger Cooper's Hawk, which has larger head, longer wings, longer and rounded tail with bolder white tip, often appears more capped; in flight, Cooper's holds wings out straighter, accentuating big head, not pushed forward like small-headed Sharp-shinned. Cf. Double-toothed Kite (p. 106). Usually quiet in winter. **STATUS:** Rare Oct–Mar; more widespread and numerous in fall migration, mainly Oct to mid-Nov. (Breeds N America to Mexico, winters to Panama.)

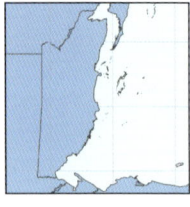

COOPER'S HAWK *Accipiter cooperi* 38–44cm male, 44–51cm female, WS 68–86cm. Rare winter migrant to wooded and semi-open habitats. Sometimes perches on roadside posts and utility poles. Flies with strong, quick, stiff wingbeats; soars on flattish wings held out relatively straight from body, long tail slightly spread to show rounded tip. Female appreciably bigger than male, can be mistaken for smaller buteos; male easily confused with smaller Sharp-shinned Hawk (which see). Cf. Double-toothed Kite (p. 106). Usually quiet in winter. **STATUS:** Rare Oct–Mar; more widespread and numerous in fall migration, mainly Oct to mid-Nov. (Breeds N America to Mexico, winters to Costa Rica.)

SMALLER 'ROADSIDE' BUTEOS

Two species that hunt mainly from perches and are often seen on roadside wires, utility poles, fence posts. Note overall plumage coloration, details of tail pattern.

GRAY HAWK *Buteo [nitidus] plagiatus* 42–46cm, WS 81–94cm. Small hawk of forest edge, semi-open habitats with taller trees, forest patches, gallery forest; not usually in rainforest. Commonly seen on roadside wires and utility poles. Soars frequently, often quite high, with wings flattish, tail usually slightly spread; frequently in kettles with vultures and other hawks. Adult distinctive, with pearly gray plumage barred below, brown eyes, bold black-and-white tail bands. Juv. best identified by size, shape, habits; whitish face with bold dark eyestripe and mustache; also note fairly long tail with pale bands slightly wavy, progressively wider toward tip, dusky barring on thighs, white uppertail coverts; in flight often appears uniformly pale below without distinct dark trailing edge to underwing of juv. Broad-winged Hawk. **SOUNDS:** Adult has drawn-out, slurred, overall descending scream, *wheeeeeu*, about 1.5 secs, not as hoarse as Roadside Hawk; juv. has huskier scream. 'Song' a distinctive series of inflected mournful whistles, often given around or before dawn, typically 3–10×, *hu-weeoo hu-weeoo....* **STATUS:** Fairly common; occasional on Ambergris Caye. (Mexico and sw. US to Costa Rica.)

ROADSIDE HAWK *Rupornis magnirostris* 33–41cm, WS 68–79cm. Conspicuous small hawk of varied habitats, from ranchland with scattered trees to humid forest interior. Typically seen perched on roadside wires and fence posts, or in low flight with rapid, stiff, rather accipiter-like wingbeats. Soars mainly in late winter–spring, when nesting, and often vocal in display flight, when can climb high: fluttering wingbeats interspersed with glides on wings raised in a strong V; at other times soars on flattish wings. Adult distinctive, with staring pale yellow eyes, gray-brown to grayish head and breast, rusty-barred belly; bright rusty primary patches show well in flight. Juv. has distinctive streaked breast but barred belly, broad pale tail bands. **SOUNDS:** Nasal, complaining, overall descending scream, *meeahh*, about 1 sec. 'Song' (mainly in display flight, less often from perch) an often persistent, fairly rapid series of clipped nasal yaps, intensifying and fading, *heh-heh-heh...*, can suggest Lineated Woodpecker. **STATUS:** Fairly common to common on mainland, uncommon on Ambergris Caye. (Mexico to S America.)

adult

SHARP-SHINNED
HAWK

1st-years

adult

1st-years

COOPER'S HAWK

adult

adult

GRAY HAWK

1st-year

adult

1st-years

adult

ROADSIDE HAWK

1st-year

adult

adult

1st-year

'AERIAL' BUTEOS Three species rarely seen perched; Short-tailed and Zone-tailed hunt in flight and are widespread over diverse habitats, Broad-winged is mainly a forest-based bird and transient migrant.

BROAD-WINGED HAWK *Buteo platypterus* 38–43cm, WS 82–92cm. Rare winter migrant to forest and edge. Mainly seen soaring in mid–late morning, often with kettles of vultures and other hawks; less often encountered perched, mainly in subcanopy and at edges. In flight, note relatively tapered wings, evenly pale flight feathers (primaries translucent when backlit) with dark trailing edge; adult shows single broad whitish band on underside of tail. At rest, note small size, relatively long wings extending well down tail. Juv. has variable dark streaking below, dark spotting (not barring) on thighs. Rare dark morph (all ages) has solidly blackish-brown body and underwing coverts. Cf. longer-tailed imm. Gray Hawk, dark Short-tailed Hawk (duskier secondaries from below). SOUNDS: High, thin, slightly tinny scream *ssiiiiiiu*, about 1 sec, from perch and in flight. STATUS: Rare to uncommon Oct–Mar, mainly in south (very rare n. of mapped range); more widespread and numerous in fall migration, mainly late Sep–Oct on s. coast; vagrant to larger cayes. (Breeds N America, winters Mexico to S America.)

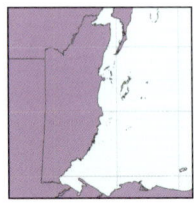

SHORT-TAILED HAWK *Buteo brachyurus* 41–46cm, WS 86–102cm. Widespread small buteo, rarely seen perched; despite the name, tail not strikingly short. Light morph and dark morph about equally common, often together in pairs. Usually seen as singles or pairs soaring over forested and semi-open habitats, including towns, mangroves, often with mid–late morning kettles of vultures and other hawks. Hunts from flight, often hanging stationary in breeze and stooping on prey. In flight, note slightly tapered wing-tips, slightly pinched-in trailing edge to wings, dusky secondaries vs. whiter primaries; soars with wings flat but distinctively curled up at tips. Dark morph juv. often checkered whitish on underparts. On perched birds, note long wings reaching to near tail tip, tawny wash on sides of neck. SOUNDS: Slightly modulated or overslurred whistled scream, *wheéeu* or *klee-ee*, usually in short series, mainly by pairs in flight. STATUS: Uncommon to fairly common. (Mexico and s. US to S America.)

ZONE-TAILED HAWK *Buteo albonotatus* 45–53cm, WS 122–137cm. Scarce migrant hawk easily passed off in flight as a Northern Turkey Vulture. Occurs in wide variety of open and wooded habitats from coastal towns and ranchland to rainforest edge. Usually seen as singles, soaring and gliding with wings in a shallow V much like Northern Turkey Vulture, with which it often associates in kettles and at roosts. Appreciably smaller than Northern Turkey Vulture, with narrower, more even-width wings; yellow cere often stands out at long range. Appreciably larger than Short-tailed Hawk, with longer wings and tail, different flight manner. On perched birds, cf. much stockier Common Black Hawk. Adult has bold white tail band below; juv. has more finely barred flight feathers, variable whitish flecking on underparts. Usually silent in Belize. STATUS: Scarce to uncommon (overlooked?) late Sep–Apr. (Mexico and sw. US to S America.)

BROAD-WINGED HAWK

1st-year

adult

1st-year

adults

dark morph

SHORT-TAILED HAWK

light morphs

adults

1st-year

dark morph

1st-year

dark morphs

ZONE-TAILED HAWK

adult

with Northern
Turkey Vulture

1st-year

adult

LARGER OPEN COUNTRY HAWKS

Three larger hawks typical of open and semi-open habitats, not in humid tropical forest; Red-tailed is mainly in Mountain Pine Ridge, Swainson's is a rare nonbreeding migrant, and White-tailed is local in savannas.

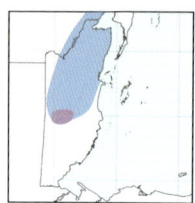

RED-TAILED HAWK *Buteo jamaicensis* 48–59cm, WS 117–137cm. Large, rather stocky hawk, mainly in Mountain Pine Ridge. Found in semi-open and open habitats with trees or utility poles for perches, forest edge; not in forested humid lowlands. Soars with wings in shallow V, glides on flattish wings; hunts from perches and in flight. Kites and hovers, mainly in open country when windy. Note thickset shape with broad, rather blunt-tipped wings, medium-length tail (1st-year has narrower wings, longer tail); light morph has dark leading edge to underwing. Plumage highly variable (dark morph rare in Belize); migrant adult has variable dark belly band. On perched birds, all except darkest morphs show distinct pale mottling on scapulars. Adult has diagnostic rusty tail. Juv. can be confusing, note size and structure, brown tail with numerous narrow dark bars, cf. Swainson's Hawk (narrower, longer wings), Short-tailed Hawk (smaller, rarely seen perched). Attains adult appearance in 2nd year. **SOUNDS:** Classic Hollywood hawk scream: drawn-out, rasping, overall descending *whieeéahrr*, mainly in flight; juv. has higher, shriller *wheéirr*, at times repeated persistently. **STATUS:** Uncommon resident in Mountain Pine Ridge and adjacent foothills; rare and sporadic elsewhere, mainly Oct–Mar in north, when small numbers of n. migrants occur; very rare in s. lowlands. (N America to Panama.)

SWAINSON'S HAWK *Buteo swainsoni* 48–56cm, WS 119–132cm. Rare nonbr. migrant. Fairly large but lightly built buteo with relatively long, narrow, tapered wings, variable plumage. Favors open country, such as ranchland, rice fields, but transients could occur over any habitat, perhaps in association with other migrating raptors. Flight buoyant and agile, soaring with wings in a shallow V. Perches readily on ground, also on low bushes, trees, utility poles. In flight, note relatively narrow, tapered wings with contrast between pale coverts and dark flight feathers; light-morph adult has dark breast band. At rest, note long wings, projecting to or just beyond tail tip, small head and bill. Uncommon dark morph has dark body contrasting with paler undertail coverts, rusty underwing coverts. Juv. has buff face and underparts, fading to whitish, with dark eyestripe and mustache. Cf. larger and bulkier White-tailed Hawk, Red-tailed Hawk. Usually silent in Belize. **STATUS:** Rare late Sep–Mar, mainly as late Sep–Nov transient on s. coast; rare and sporadic (increasing?) in winter, when possible anywhere in open country. (Breeds N America to Mexico, winters Mexico to S America.)

WHITE-TAILED HAWK *Geranoaetus albicaudatus* 48–59cm, WS 124–137cm. Large hawk local in savanna and ranchland with scattered trees and bushes; ranges to adjacent wetlands. Perches mainly on low posts and bushes. Soars and glides with wings in distinct shallow V; at times kites and hovers, especially over fields being burned; hunts mainly in flight. Note long, broad-based, and strongly tapered wings, short tail (1st-year has distinctly narrower wings, longer tail); wing-tips project past tail at rest, especially on adult. Adult distinctive; 1st-year variable, most are extensively dark below, a few mostly whitish below with limited dark streaking, cf. Red-tailed Hawk; 2nd-year resembles adult but darker above, with dark hood, duller tail pattern, variable narrow dark barring on belly. Attains adult appearance in 3rd year. **SOUNDS:** Mostly quiet away from nest. Slightly hoarse screams, often couplets in short series, *whee-ah whee-ah....* **STATUS:** Uncommon and local, mainly in north; rare and sporadic s. of mapped range (Mexico and sw. US to S America.)

RED-TAILED HAWK

1st-years

1st-year

adults

dark morphs

adults

SWAINSON'S HAWK

1st-year

adult

adults

1st-year

WHITE-TAILED HAWK

1st-year

adults

1st-year

GENUS *BUTEOGALLUS*

(3 species) Large, relatively broad-winged, and short-tailed hawks with long yellow legs that may recall chickens (hence, *Buteogallus*). Found mostly in forested habitats, often near water. Frequently soar in mid–late morning. Note structure, tail pattern, and voice.

COMMON BLACK HAWK *Buteogallus anthracinus* 46–53cm, WS 109–127cm. Fairly large, heavily built hawk of marshes, mangroves, forested and semi-open areas near water, including towns, villages. Often seen perched overlooking water; can be very confiding. Plunges feet first to snatch fish; runs on beaches and mudflats hunting crabs. Soars regularly, mainly in mid–late morning, wings flattish, tail spread. Note broad wings, short tail (appreciably longer on juv.), long yellow legs. Adult has single white tail band, often shows whitish panel across base of outer primaries. Main confusion risk is Great Black Hawk (see that species for details). Common Black often mistaken for rare Solitary Eagle, which is much larger, paler slaty gray overall, with thicker legs and an even shorter tail—in flight, toes of Solitary Eagle project almost to tail tip. Also cf. perched adult Zone-tailed Hawk; juv. Gray Hawk plumage suggests juv. Common Black but note structure. Attains adult appearance in 2nd year. Some adults have underside of flight feathers variably suffused rusty. **SOUNDS:** Often calls when soaring and in display. Varied series of loud ringing whistles, often intensifying and then fading quickly, *yih yih yih yeep Yeep YEEP YEEP yeep yih-yih-yih* and variations. **STATUS:** Fairly common, mainly along coast (including inshore cayes) and major rivers, scarce inland in upland areas. (Mexico and sw. US to n. S America.)

GREAT BLACK HAWK *Buteogallus urubitinga* 51–61cm, WS 120–137cm. Large, broad-winged hawk of forested and semi-open areas, marshes. Habits much like Common Black Hawk, but usually less confiding. Soars with wings flattish, tail rarely spread widely; often dangles long legs. Less compact than Common, with longer neck, longer legs, longer tail that projects noticeably past tail tip on perched birds (especially juv.); also note voice. Adult Great Black has less extensive, often duller yellow at base of bill, coarse white barring on thighs. In flight, wings of adult Great bulge strongly on secondaries, feet project past white tail band; 2nd white tail band of Great hard to see from below but obvious from above, when can look like white tail split by black band. Juv. Great has paler head than Common, without thick dark mustache, longer tail has numerous narrow dark bars, broad dark distal band, vs. fewer and broader dark bars of Common; uppertail coverts mostly white (mostly dark on Common). 2nd-year Great Black resembles juv. but can show dark mustache and has coarse tail bars, suggesting Common; note structural differences, mostly white uppertail coverts. Attains adult appearance in 3rd year. Also cf. Crane Hawk, Solitary Eagle. **SOUNDS:** High, piercing, drawn-out wailing whistle, perched and in flight from both adult and juv., 2–5 secs. 'Song' given in flight carries well: short, overslurred, piping whistles in prolonged rapid series, *whi' pih-pih-pih....* **STATUS:** Uncommon, including upland areas. (Mexico to S America.)

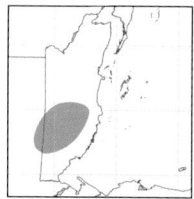

SOLITARY EAGLE *Buteogallus solitarius* 64–79cm, WS 152–188cm. Very large, rare eagle of evergreen and pine forest in foothills and highlands. Soars mainly in mid–late morning. Adult told from smaller Common Black Hawk (which can appear disconcertingly large thanks to its shape and slow wingbeats) by paler, slaty-gray plumage, relatively shorter tail, and thicker legs; in flight, toes project almost to tail tip, vs. projecting just into white tail band on Common Black. Juv. has distinctive plumage, with large dark patches at sides of breast, dark thighs, and overall plain underside to flight feathers, with no distinct tail bars. 2nd-year resembles adult but browner overall, with sparse buff streaks on head, body, underwing coverts; tail pattern like adult but broad median band pale gray, not white. Probably attains adult appearance in 3rd year. **SOUNDS:** Melancholy drawn-out whistle, 1–2 secs, lower and less piercing than Great Black Hawk. In flight, powerful whistled screams in slightly speeding and slowing series, *whieh-whieh...*, cadence reminiscent of Common Black Hawk but notes stronger, lower-pitched. **STATUS:** Rare and local in Maya Mts. (Mexico to S America.)

COMMON BLACK HAWK

1st-year

adults

dorsal

1st-year

GREAT BLACK HAWK

1st-year

adults

dorsal

1st-year

SOLITARY EAGLE

1st-year

adults

dorsal

1st-year

MISCELLANEOUS TROPICAL HAWKS Snail Kite and Black-collared Hawk are birds of wetlands and lakes; Crane Hawk is often near water but also occurs in forested and semi-open habitats far from water.

CRANE HAWK *Geranospiza caerulescens* 46–54cm, WS 92–105cm. Distinctive, rather lanky, small-headed hawk of varied semi-open and forested habitats, from mangroves and ranchland with scattered trees to coconut plantations and rainforest; often near water. Clambers around in trees, using its long, double-jointed legs to reach into crevices and cavities in search of prey, flapping its wings for balance. Wingbeats rather loose, slow, and floppy, suggesting a larger bird; soars on flattish wings, mainly for short periods in mid-morning. In display flight, flap-flap-flap-glide progression over canopy interrupted by brief climbs with quicker loose flaps, then glides back to original level. Note very long red legs, small head with ruby-red eyes and gray cere (no yellow in face, cf. black hawks), long tail with 2 white bands. In flight shows variable band of white spots across primaries (can be hard to see). Juv. has grizzled pale face, amber eyes, red-orange legs, variable pale mottling below; attains adult plumage in 2nd year. **SOUNDS:** Slightly mournful, overall descending whistled scream, *WHéeooo*. **STATUS:** Uncommon, most frequently encountered in the north. (Mexico to S America.)

SNAIL KITE *Rostrhamus sociabilis* 45–51cm, WS 104–119cm. Distinctive, broad-winged raptor of wetlands and lakes with large snails, virtually its sole food source. Mostly seen perched overlooking water or quartering over water in unhurried flight with measured wingbeats and glides on distinctively arched wings, dropping to snatch snails with its feet. Can also be seen high overhead in any habitat, sometimes in small groups, when moving between wetlands or searching for new sites that have not dried up. Usually breeds and roosts communally. On perched birds note slender bill with long hook, reddish eyes, long wings projecting to or beyond tail tip, fairly short legs. On flying birds note flight manner, gliding and soaring on arched wings, and large white base to fairly short squared tail. Juv. has buff face and underparts, fading to whitish, dark eyes, yellowish legs; adult female and 2nd-year male have mostly dark underparts, pale face; adult male dark slaty gray overall with deep red eyes, salmon-orange legs. Attains adult plumage in 3rd year. **SOUNDS:** Stuttering, harsh creaky cackle, 1–2 secs; varied low rasping calls, including a shrieky *shiéhr*. **STATUS:** Fairly common to common locally in north, uncommon to rare and sporadic in south; wandering birds might show up anywhere. (Mexico and se. US to S America.)

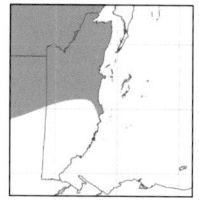

BLACK-COLLARED HAWK *Busarellus nigricollis* 46–56cm, WS 114–134cm. Distinctive, stocky, very broad-winged hawk of freshwater wetlands, lakes, slow-moving rivers, forest ponds; mainly eats fish, hunted from perches. Rather sluggish; often perches on low posts, at times on ground, also in trees overlooking water. Wingbeats slow and fairly deep, soars on flattish wings with tips often curled up. Adult essentially unmistakable, bright orangey with whitish head; black collar not always easy to see. On juv., note habits, stocky shape with short tail, ghosting of adult plumage pattern, grayish cere, pale legs; 2nd-year like adult but underparts flecked whitish and dark, thighs barred dusky. Attains adult plumage in 3rd year. **SOUNDS:** Usually quiet. Occasional hoarse rasps and a slightly piercing, slurred screaming *hileeee*. **STATUS:** Uncommon and local. (Mexico to S America.)

1st-year CRANE HAWK adults

SNAIL KITE

1st-year female males

1st-year

BLACK-COLLARED HAWK

1st-year adults

1st-year

*NORTHERN WHITE HAWK *Pseudastur [albicollis] ghiesbreghti* 48–56cm, WS 114–132cm. Breath-taking, essentially unmistakable snow-white hawk of rainforest and edge; at a distance cf. adult King Vulture. Singles and pairs soar low over canopy in mid–late morning, often revealed by loud screaming calls. Wingbeats rather slow and floppy; soars on flattish wings. Perches mainly in subcanopy, where easily overlooked. Juv. has more extensive black on wings and tail; 2nd-year like adult but with some black mottling on bases of secondaries, more extensive black in wing-tip; like adult in 3rd year. **SOUNDS:** Drawn-out, husky, slightly overslurred scream, 1–1.5 secs, typically repeated a few times; suggests a pig squealing. **STATUS:** Uncommon to fairly common, especially in hilly country. (Mexico to nw. S America.)

TROPICAL FOREST RAPTORS IN FLIGHT
Various forest species soar over the canopy and adjacent edges, especially in mid–late morning (mainly 9–11.30 am, as thermals develop; smaller species earlier, larger species mainly after 10 am); many species call and make display flights that involve stooping and other aerobatic maneuvers. Opposite we show the species and plumages most likely to be seen (not strictly to scale, as size is difficult to judge on lone birds); also cf. Black and King Vultures (p. 90).

SHORT-TAILED HAWK (p. 96) Small; often high and at times kites in wind; soars with wings flat, tips curled up. Light and dark morphs readily occur together, often in kettles with vultures.

HOOK-BILLED KITE (p. 106) Medium-size, with highly variable plumage; floppy wingbeats can make it look larger. Note rounded wings with broad secondaries, fingered primaries with bold checkering; wide tail bands; usually silent. Black morph rare, but adult male often appears dark overall with bold white checkering in primaries. Unlike most other hawks, imms. often seen soaring, at times in small groups with adults.

DOUBLE-TOOTHED KITE (p. 106) Small, accipiter-like; tail often held closed, showing white puffy tail coverts; soars with wings slightly bowed, often in pairs; high thin whistles distinctive but easily missed.

PLUMBEOUS KITE (p. 114) Medium-size, pointed wings, leisurely soaring and flapping, catching insects.

BLACK-AND-WHITE HAWK-EAGLE (p. 108) Rare; fairly large with relatively tapered wings, short tail; often soars high and hunts by stooping on prey; note white head and black mask, white leading edge to upperwing striking from above; calls infrequently, unlike other hawk-eagles.

GRAY-HEADED KITE (p. 106) Medium-size, small-headed; floppy wingbeats; bold black wing linings contrast with white body (imm. rarely seen soaring); often silent.

ORNATE HAWK-EAGLE (p. 108) Fairly large, with broad secondaries, rounded wings; often soars high and usually detected and identified by loud whistled 'song' that carries far; often looks rather plain, with distinctive head and body plumage only apparent at closer range.

BLACK HAWK-EAGLE (p. 108) Fairly large, with broad secondaries, rounded wings; often soars high and usually detected and identified by loud whistled 'song' that carries far; often looks black overall with white checkering only obvious on primaries.

CRANE HAWK (p. 102) Medium-size, but floppy flight can make it look larger; note long tail with 2 white bands, long reddish legs, small head (no yellow in face); band of white spots in primaries can be hard to see; usually silent.

GREAT BLACK HAWK (p. 100) Fairly large, broad-winged (slow wingbeats often make it look much larger; regularly misidentified as much rarer Solitary Eagle). Tail slightly longer than Common Black, with 2 white bands (hard to see from below), longer yellow legs (often dangled); prolonged, high piping slow trill carries well, distinctive.

COMMON BLACK HAWK (p. 100) Fairly large (slow wingbeats often make it look much larger; regularly misidentified as much rarer Solitary Eagle), with broad wings, short tail; note single white tail band, yellow legs/feet reach just into tail band; often vocal, giving short series of loud piping whistles.

SOLITARY EAGLE (p. 100) Very large, broad-winged, short-tailed (but cf. black hawks); note single white tail band, long yellow legs/feet reach near tail tip.

1st-year

NORTHERN WHITE HAWK

adults

1st-year

Short-tailed Hawks

Hook-billed Kites

Double-toothed Kite

Black-and-white Hawk-Eagle

Gray-headed Kite

Plumbeous Kite

Ornate Hawk-Eagle

Hook-billed Kites

Black Hawk-Eagle

Common Black Hawk

Crane Hawk

Great Black Hawk

Solitary Eagle

ROUNDED-WINGED FOREST KITES
Small to medium-size forest-based raptors; imms. in particular have rather variable plumage. Often seen soaring, especially in mid–late morning; relatively slow floppy wingbeats of Hook-billed and Gray-headed can make them seem larger, cf. hawk-eagles.

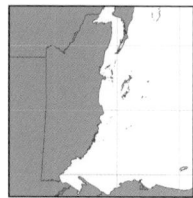

DOUBLE-TOOTHED KITE *Harpagus bidentatus* 32–36cm, 64–74cm. Small, rather accipiter-like kite of rainforest and semi-deciduous forest. Often one of the first rainforest raptors to soar in mid-morning, as singles or pairs, at times stooping in display flights. Typical flight on distinctively bowed wings, with tail closed and puffy white undertail coverts flared; at other times soars on flattish wings, tail slightly spread, much like an accipiter. Perches quietly in subcanopy and at edges; at times accompanies bands of monkeys, which flush insect prey from foliage. In flight, note slightly paddle-shaped wings (longer than accipiters), whitish underwing coverts, bold dark barring on primaries; on adult, rusty underparts contrast with white underwing coverts. On perched birds, note greenish-yellow facial skin, dark median throat stripe; double notch on cutting edge of bill (double 'tooth') rarely visible. Cf. Cooper's Hawk, larger and bulkier Broad-winged Hawk. **SOUNDS:** High thin whistles given in flight by displaying birds and pairs, *tseéu-ip*, and *tseéu tsee-u*, faster *tsip tsí-yiip*, and variations; high, slightly shrill whistled *shiiep* from perched birds. **STATUS:** Fairly common to uncommon; occasional on Ambergris Caye. Apparent migrants in s. Belize (mid-Oct to mid-Dec) suggest some seasonal movement. (Mexico to S America.)

HOOK-BILLED KITE *Chondrohierax uncinatus* 38–46cm, WS 81–94cm. Medium-size, broad-winged kite with large bill, bright facial skin, variable plumage. Found in varied forested habitats where feeds mainly on land snails. Soars often, but rarely for long periods, mainly in mid–late morning, at times in small groups. Wingbeats rather loose and floppy, often suggesting a larger bird. Soars with wings flattish, tail slightly to widely spread. The floppy wingbeats and paddle-shaped wings, with pinched-in bases and spread-fingered hands, separate Hook-billed Kite from superficially similar buteos, which have stiffer wingbeats, different tail patterns, etc. Cf. larger Ornate Hawk-Eagle (plainer underwings), imm. Gray-headed Kite (broader tail bands, whiter head). On perched birds, note big bill, bright facial skin, small feet. Adult male barred gray below (can appear dark against the sky, with bold white checkering in primaries), female barred rusty; 1st-year plain whitish below or with variable dark barring. Scarce dark morph (sexes similar) blackish overall with 1–2 broad pale tail bands. **SOUNDS:** Infrequently heard. From perch, a rapid, overslurred, slightly chuckling chatter of about 10–20 notes, *weh keh-eh-eh…*, 1–1.5 secs; may suggest a woodpecker. **STATUS:** Uncommon but widespread presumed resident; locally numerous transient on s. coast, mid-Oct to Nov (smaller numbers into mid-Dec), with seasonal fall totals up to 5000 or so birds. (Mexico to S America.)

GRAY-HEADED KITE *Leptodon cayanensis* 45–53cm, WS 92–107cm. Fairly large, broad-winged kite of rainforest, semi-deciduous forest and edge, especially along rivers and near water. Soars on flattish wings, tail rarely spread widely; wingbeats typically languid and easy. In display, interrupts glides with short steep climbs powered by deep, quick, floppy wingbeats followed by a brief glide down with wings held in deep V. Adult distinctive, with soft gray head, dark eyes, blue-gray cere and feet; in flight, black underwing coverts contrast with white body. Imm. plumages variable, typically plain whitish below but sometimes with variable dark streaking; note long, broad, slightly rounded tail with broad pale bands, small feet, brownish eyes. Cf. Hook-billed Kite, Black-and-white Hawk-Eagle. **SOUNDS:** From perch and in flight display, a fairly rapid, steady, at times rather prolonged series of hollow laughing clucks, 4–5 notes/sec, *kyuh-kyuh…*, or *keh-keh…*, usually 5–12 secs; recalls Lineated Woodpecker but lower, less nasal, with shorter notes. Overslurred to slightly inflected wailing *meeaowh*, rather mammal-like, about 1 sec. **STATUS:** Uncommon, mainly in lowlands. (Mexico to S America.)

DOUBLE-TOOTHED KITE

1st-year

adults

1st-years

adult

HOOK-BILLED KITE

dark morph

males

female

1st-year

female

1st-year

1st-year

GRAY-HEADED KITE

adults

1st-years

1st-year

HAWK-EAGLES (GENUS *SPIZAETUS*)
Fairly large crested raptors of humid forest. Most often seen in flight, especially between 10 am and noon, when morning thermals are strongest before cloud build-up. Like many tropical raptors, juvs. seen very infrequently.

BLACK-AND-WHITE HAWK-EAGLE *Spizaetus melanoleucus* 51–64cm, WS 117–142cm. Scarce eagle of rainforest, less often semi-deciduous forest. Soars and glides on flattish wings, tips sometimes curled up; tail slightly spread when soaring; hunts by stooping from considerable heights. Relatively long-winged and short-tailed compared with other hawk-eagles, suggesting much smaller Short-tailed Hawk. Also note white head with black lores and striking, diagnostic white leading edge to upperwing, which shows well at long range when birds bank and wheel; short, spiky black crest apparent with good views. Cf. Short-tailed Hawk, imm. Gray-headed Kite, imm. Ornate Hawk-Eagle. 1st-year similar to adult but browner above, tail bars narrower (4–5 dark bars vs. 3–4 on adult); attains adult appearance in 2nd year. SOUNDS: Calls infrequently, unlike other hawk-eagles. Piercing to piping whistles in flight suggest Ornate Hawk-Eagle in quality, but have steadier rhythm: even-paced to slightly accelerating series may end with a more emphatic, at times disyllabic whistle, *whee whi-whi-whi-whi whee-eer*, and variations. STATUS: Scarce to uncommon, especially in hilly areas. (Mexico to S America.)

ORNATE HAWK-EAGLE *Spizaetus ornatus* 56–69cm, WS 117–142cm. Spectacular, fairly large eagle of rainforest, semi-deciduous forest. Soars often in mid–late morning, when usually detected by far-carrying whistled calls. Soars on flattish wings, tail slightly spread; wingbeats deep and powerful. In display, climbs with deep floppy wingbeats and stoops with wings closed, almost somersaulting at times. Perches mainly in subcanopy, where can be confiding. Perched adult unmistakable, but more often seen high overhead, when looks rather uneventful and best identified by voice: note fairly plain underwings, bold tail banding, white bib contrasting with rusty head sides. 1st-year rarely seen but distinctive, with barred flanks, white head with blank face, long spiky crest; 2nd-year like adult but averages duller and messier; attains adult appearance in 3rd year. Cf. Hook-billed Kite, Gray-headed Kite, Black-and-white Hawk-Eagle. SOUNDS: Far-carrying piping whistles, often repeated tirelessly when soaring, usually with a pause after 1st note and overall a distinctive, slightly stuttering cadence: *whi, whee-whee-wheep, wh, whi, whee-whee-wheep…*, and variations. Unlike Black Hawk-Eagle, introductory series more hurried and last note not drawn out. Perched juv. has loud clear whistle, repeated, *wheeeu*. STATUS: Uncommon, mainly away from coast and in hilly areas. (Mexico to S America.)

BLACK HAWK-EAGLE *Spizaetus tyrannus* 61–74cm, WS 127–155cm. Spectacular, fairly large eagle of rainforest and edge, semi-deciduous forest, plantations; more tolerant of cut-over and second-growth forest patches than other hawk-eagles. Soars often in mid–late morning, when usually detected by far-carrying whistled calls. Soars on flattish wings, tail slightly spread; wingbeats deep and powerful. In display, soars with tail closed, wing-tips quivering. Perches mainly in subcanopy and at edges. Perched adult distinctive, with golden eyes, bushy white-based crest, boldly barred leggings, long tail. In flight often looks black overall against the sky, with bold barring only in primaries. 1st-year distinctive, with bushy crest, broad black mask, heavily barred underparts; 2nd-year like adult but averages messier, with scattered whitish flecking on head and underparts; attains adult appearance in 3rd year. Cf. dark Hook-billed Kite, Ornate Hawk-Eagle. SOUNDS: Far-carrying piping whistles, often repeated tirelessly when soaring. Loud, clear, overslurred whistles, often preceded by 1 or more shorter whistles not always audible at a distance, *wheéoo* or *wh-wheéoo*; also *whi-whi whi-wh-wh-wheéoo* and variations. Imm. has steady ringing whistles, *whee whee…*, perched and in flight; perched adult may give single *weeoo*. STATUS: Uncommon to fairly common; the most frequently encountered hawk-eagle in Belize. (Mexico to S America.)

BLACK-AND-WHITE HAWK-EAGLE

adults

ORNATE HAWK-EAGLE

adults

1st-year

1st-year

BLACK HAWK-EAGLE

adults

1st-year

1st-year

110

BIG RAINFOREST EAGLES
Very large to massive crested raptors of tropical forest; unlike hawk-eagles, do not soar and very rarely seen—in our experience you have as much or more chance of seeing a jaguar.

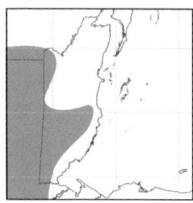

CRESTED EAGLE *Morphnus guianensis* 71–87cm, WS 152–185cm. Very rarely encountered. Very large eagle of remote rainforest and edge. Hunts from canopy and subcanopy perches, preying on mammals and larger birds. Rarely soars and hence overlooked easily. Most likely to be seen perched along waterways and sunning itself in early or mid-morning on emergent canopy snags; can be confiding. Cf. larger, bulkier, and more powerful Harpy Eagle, which has shaggier, forked crest. Hawk-Eagles are appreciably smaller, with different plumage patterns, feathered legs, staring golden-yellow eyes (eyes brown to dull yellowish on Crested Eagle). Adult Crested variable, typically has pale rusty to blackish barring on underparts; exceptionally all-dark below. Attains adult plumage in about 3 years. **SOUNDS:** Perched adult has short series of full-bodied, slightly overslurred whistles, introduced by 1–2 shorter notes, *whi whi wheeéooo wheeéooo wheeéooo*; longer, at times persistent series of short overslurred whistles, *whi wheeé wheeé wheeé…*; and plaintive, slightly rising, clear-toned whistle, 1–1.5 sec, higher, less mournful than Harpy Eagle. **STATUS:** Very rare and local. (Mexico to S America.)

HARPY EAGLE *Harpia harpyja* 87–107cm, WS 183–224cm. Very rarely encountered. Huge powerful eagle of southern rainforest and edge. Hunts from canopy and subcanopy perches, preying on mammals and larger birds. Does not soar, and hence overlooked easily. Most likely to be seen perched along waterways and sunning itself in early or mid-morning on emergent canopy snags; can be confiding. Flights mainly short and low over canopy or across rivers. Only possible confusion species is slightly smaller and equally rare Crested Eagle. Crested is less heavily built (but can still appear huge), overall slimmer and longer-tailed. Crested has smaller bill, less massive legs, and erectile crest is single-pointed. In flight, Crested shows heavier dark barring across secondaries and inner primaries; adult Crested has gray hood (without Harpy's black chest patch) and has unmarked whitish underwing coverts. Harpy also has broader tail banding relative to similar age Crested, and attains adult plumage in about 4 years. **SOUNDS:** Mainly vocal around nest. Adult gives mournful, slurred wailing whistles, 1–1.5 secs; imm. has higher, shriller, drawn-out whistles. **STATUS:** Very rare and local. (Mexico to S America.)

1st-year

adults

Crested Eagle

adult

1st-year

Harpy Eagle

adults

1st-year

1st-year

BICOLORED HAWK, FOREST FALCONS
Rarely seen tropical forest species that do not soar. Forest falcons, however, are often heard, especially early and late in the day.

BICOLORED HAWK *Accipiter bicolor* male 33–38cm, female 43–48cm, WS 61–79cm. Rarely seen hawk of humid forest. Most likely to be encountered perched at edges or on exposed snags, especially early to mid-morning. Flight rapid and direct with quick wingbeats, brief glides; not known to soar. Adult distinctive, with plain pale gray to gray underparts, staring amber eyes; rusty thighs often difficult to see. 1st-year has rich buff to whitish underparts, most likely to be confused with forest falcons, which share similar habits: note staring pale yellow eyes of Bicolored, plus subtly different extent and pattern of bare facial skin; forest falcons have more strongly graduated tails with narrower, more contrasting whitish bars. Attains adult appearance in 2nd year. **SOUNDS:** Much like Cooper's Hawk of North America, and rarely vocal away from nest. Adult has variably paced series of barking clucks, 4–8 notes/sec, *keh-keh-keh…*, and downslurred, rough snarling mew. Juv. gives overall descending wheezy scream. **STATUS:** Scarce to uncommon. (Mexico to S America.)

BARRED FOREST FALCON *Micrastur ruficollis* 33–38cm, WS 49–59cm. Heard far more often than seen; a retiring, fairly small raptor of rainforest, less often semi-deciduous forest. Usually calls from concealed perch at low to mid-levels in forest, when elusive; at other times tends to perch quietly inside forest, often fairly low, and can be confiding; sometimes attends army ant swarms to prey on small birds. Flight fast and direct, accipiter-like; does not soar. Adult distinctive, with bright yellow face, finely barred underparts. 1st-year much smaller and less lanky than Collared Forest Falcon, with brighter facial skin, less strongly graduated tail. Cf. Bicolored Hawk. Attains adult appearance in 2nd year. **SOUNDS:** Overslurred yapping bark, repeated steadily, *káah, káah,…* 10 barks/12–18 secs, and more hesitant series mainly around dawn; also short, slightly descending series, usually 5–8 notes, with slightly laughing cadence, *kah, kaah-kaah-kaah-kah*; cf. quieter but rather similar-sounding calls of Plain Antvireo. **STATUS:** Uncommon; most numerous in hilly country. (Mexico to S America.)

COLLARED FOREST FALCON *Micrastur semitorquatus* Male 53–56cm, female 61–64cm, WS 76–94cm. Heard far more often than seen; a spectacular, rather large, lanky, and long-tailed raptor of forested habitats; also ranges to semi-open areas with hedgerows and forest patches. Usually calls from high perch in subcanopy, and can be elusive. At other times ranges low to high, mostly perching quietly inside forest; sometimes attends army ant swarms. Flight direct, with fairly quick loose wingbeats and short glides; does not soar. Distinctive when seen, with dark eyes, dark cheek crescent, very long graduated tail, long legs; overhead, note boldly barred and checkered underside to flight feathers. Cf. 1st-year of smaller Bicolored Hawk. Most likely to be confused by sound with Laughing Falcon. Adults typically white below, rarely buff; dark morph very rare. Attains adult appearance in 2nd year. **SOUNDS:** Far-carrying, hollow, slightly overslurred *cowh* or *owhh*, in short series or repeated steadily but not hurriedly, mainly early and late in day; rarely faster than 1 note every 1.5–2 secs; at times breaks into an accelerating then slowing laugh of about 12–20 clucks, ending with a pause and final plaintive note, *hoh-hoh-hoh…, owh*. Laughing Falcon has shorter, higher, typically faster-paced notes with, dare we say it, a more laughing quality (by comparison, forest falcon sounds as if it's being hit, *owh…*). Also quiet single notes on occasion. Juv. has a more plaintive *mehow* or *kyeow*, and quiet reedy chippering in alarm. **STATUS:** Uncommon to fairly common. (Mexico to S America.)

BICOLORED HAWK

1st-years

adult

BARRED
FOREST FALCON

1st-years

adult

1st-years

COLLARED
FOREST FALCON

adults

dark
morph

114

POINTED-WINGED KITES Four species that hunt in flight, White-tailed while hovering like a kestrel, the others while soaring and gliding like giant swallows.

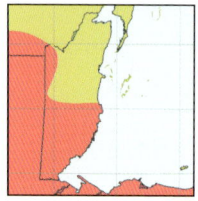

SWALLOW-TAILED KITE *Elanoides forficatus* 56–61cm, WS 117–135cm. Summer migrant. Spectacular aerial raptor with long pointed wings and very long, deeply forked tail; no similar species in Belize. Breeds in open humid forest and pine savanna, often near water. Migrants occur widely in any open and forested habitats. Flight graceful and buoyant with deep easy wingbeats and leisurely soaring, snatching insects from canopy and on the wing; soars on flattish wings with tips curled up. Often in pairs or small groups; post-breeding and migrant flocks sometimes 100+ birds. 1st-year has shorter tail than adult, white tips to primaries and primary coverts. Attains adult appearance in 2nd year. SOUNDS: Shrill piping and ringing whistles, mainly in flight, at times in rapid yelping series, accelerating and slowing, *kleeh kleeh-kleeh-kleep* and variations. STATUS: Fairly common locally Feb–Aug, especially in upland areas; local post-breeding gatherings late Jul–Aug in north may be away from breeding areas; more widespread as a transient, mainly Aug–Sep (stragglers into Oct), Feb–Mar. (Breeds N America to S America, winters S America.)

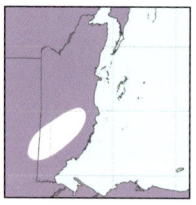

WHITE-TAILED KITE *Elanus leucurus* 38–41cm, WS 89–99cm. Distinctive elegant raptor with pointed wings, fairly long whitish tail, black shoulders. Favors open country with scattered trees, savanna, marshes, ranchland. No really similar species, but cf. male Northern Harrier. Flies with easy wingbeats, wings held in a shallow V during glides and infrequent soaring. Often hovers and perches on roadside wires, like a kestrel. Juv. has rusty wash to breast, browner upperparts with whitish feather tips. Attains adult appearance in 2nd year. SOUNDS: Varied low rasping notes, at times paired with rising whistles; series of short, downslurred whistles, *hüw, hüw…*, recalling Osprey. STATUS: Uncommon to fairly common, spreading with deforestation; occasional visitor to Ambergris Caye. (N America to S America.)

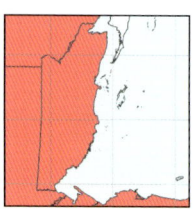

PLUMBEOUS KITE *Ictinia plumbea* 33–37cm, 84–94cm. Summer migrant. Dark aerial kite of rainforest and edge, semi-deciduous forest, gallery forest, often near rivers. Spends much time soaring and gliding; catches insects on the wing. Soars on flattish wings, the tips often curled up, tail slightly spread; wingbeats languid and smooth. Often perches high in trees adjacent to clearings and rivers. Main confusion risk is with paler Mississippi Kite (limited seasonal overlap); also cf. falcons. 1st-summer has mostly adult-like head and body, retains juv. wings and tail. Attains adult appearance in 2nd year. SOUNDS: From perch and in flight, a high, plaintive, downslurred whistle, typically preceded by a short overslurred note, *si-hiiew*; short piping whistles. STATUS: Uncommon to locally fairly common mid-Feb to Aug (very rarely from Jan and into Sep). (Breeds Mexico to S America, winters S America.)

MISSISSIPPI KITE *Ictinia mississippiensis* 34–38cm, WS 84–94cm. Transient migrant that could occur over any habitat, mainly forest and edge in coastal lowlands. Soars on flattish wings, the tips often curled up, tail slightly spread; gliding shape suggests a falcon, but wingbeats languid and smooth. Confusion most likely with darker Plumbeous Kite (little seasonal overlap); also cf. Peregrine Falcon. Plumbeous Kite has rusty primary flashes (can be hard to see); adult has white tail bands, lacks pale secondary panel on upperwing; juv. has sparser dark brown (not rusty) streaking below. 1st-summer Mississippi has mostly adult-like head and body, retains juv. wings and tail. Attains adult appearance in 2nd year. SOUNDS: Usually silent; calls much like Plumbeous Kite, average higher-pitched. STATUS: Uncommon to briefly fairly common transient Sep–Nov in south (rarely stragglers into Dec), rare and sporadic n. of mapped range, mainly juvs. in late Sep–Oct; scarce and irregular mid-Mar to Apr. (Breeds N America, winters S America.)

SWALLOW-TAILED KITE

adults

WHITE-TAILED KITE

juv.

adult

PLUMBEOUS KITE

juvs.

adults

adults

adults

juvs.

MISSISSIPPI KITE

adult

FALCONS (FALCONIDAE; 10 SPECIES; including forest falcons on p. 112, caracara and Laughing Falcon, p. 92). Fairly diverse, worldwide family of raptors; genetic studies indicate falcons share a common ancestor with parrots, but they have long been grouped with hawks as 'birds of prey,' as done here for ID purposes. Ages differ, sexes similar or different (female often noticeably larger); like adult in 2nd year.

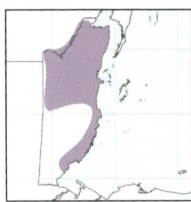

APLOMADO FALCON *Falco femoralis* 38–46cm, WS 81–93cm. Attractive, lightly built, and long-tailed falcon of pine savanna, marshes, other open habitats with scattered trees. Perches on trees, low shrubs, roadside wires and posts. Flight smooth and powerful, dashing after small birds, hunting over burning fields, and soaring in a leisurely manner, snatching insects in flight. Distinctive, with very long tail, bold white eyebrow, black breast band, cinnamon belly; also note white trailing edge to wings, narrow white tail bars. Cf. larger and stockier Peregrine Falcon. Juv. has browner upperparts, heavier dark streaking on chest, buff suffusion to face and breast in fresh plumage; blue-gray cere and eyering become yellow like adult over 1st year; attains adult appearance in 2nd year. **SOUNDS:** Fairly high screaming calls, at times in persistent rapid series, *kieh-kieh…*; single sharp *keeh* or *kiih*. **STATUS:** Uncommon to fairly common, but often rather local. (Mexico and sw. US to S America.)

BAT FALCON *Falco rufigularis* 23–28cm, WS 61–73cm. Handsome small falcon of humid forest and edge, adjacent clearings, semi-open areas with scattered trees and forest patches; locally in towns, often at Maya ruin sites. Typically perches conspicuously on bare snags, antennas, buildings, often in pairs, when size difference between sexes readily apparent (female larger). Flight fast and powerful, chasing birds, bats, larger insects, and dive-bombing larger raptors; soars on flattish wings, at times high overhead when might be mistaken for White-collared Swift. Also cf. rare and much larger Orange-breasted Falcon, migrant Merlin. Note Bat Falcon's small size, overall dark plumage with whitish throat, rusty thighs. Throat and neck sides variably suffused buff, on some birds strongly orangey, inviting confusion with Orange-breasted Falcon. 1st-year duller overall, with dark breast streaks, barring on thighs and undertail coverts. **SOUNDS:** Penetrating, at times persistent screaming from perch and in flight: slower-paced *krieh krieh….* and rapid *hew-hew…*; single sharp *kik*. **STATUS:** Fairly common to uncommon. (Mexico to S America.)

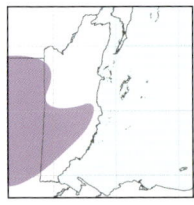

ORANGE-BREASTED FALCON *Falco deiroleucus* 35–41cm, WS 77–90cm. Large, powerfully built falcon of rainforest and edge, adjacent pine forest, often near cliffs. Singles and pairs often perch on prominent snags, like Bat Falcon. Flight powerful and direct, hunts birds in flight. Note large size; heavily built structure with big bill and very large feet relative to Bat Falcon; white throat contrasting with orange breast. Resembles Bat Falcon in plumage, but size, structure, and voice more like Peregrine Falcon. Fresh juv. plumage has throat, neck sides, and underparts suffused buffy cinnamon, with dark breast streaks, barring on thighs and undertail coverts; fades to pale buff over 1st year. **SOUNDS:** Hard, often insistent screaming *kyowh-kyowh…*, or *kyah-kyah…*, may suggest a fast-paced Brown Jay, but more clipped. Also single, more barking screams, *kyow* or *kyowh*. **STATUS:** Rare and local, mainly at a few sites in Cayo, including Hidden Valley Falls; exceptional wanderer in s. lowlands away from known territories. (Guatemala, formerly Mexico, to S America.)

APLOMADO FALCON

1st-year

adults

BAT FALCON

variation

1st-year

adults

cf. White-collared Swift

ORANGE-BREASTED FALCON

faded 1st-year

adults

AMERICAN KESTREL *Falco sparverius* 26–29cm, WS 61–65cm. Attractive, lightly built small migrant falcon of varied open and semi-open habitats with trees and scattered bushes, especially farmland, but also pine savanna, rural areas; often perches on roadside wires and utility poles. Wingbeats typically rather loose and floppy, not as strong, clipped, and purposeful as Merlin; soars on flattish wings, often hangs in wind and hovers. Distinctive: all plumages have complex head pattern, long rusty tail, overall pale underwings. Male has bluish wings, bright rusty tail with black distal band; female has rusty brown upperparts and rusty tail barred blackish. Juv. male has whiter breast, heavier black spotting than adult. **SOUNDS:** High shrill screaming *kyieh-kyieh…* in interactions; usually in short bursts, repeated. **STATUS:** Uncommon Oct–Mar (a few from mid-Sep and into Apr), mainly in north. (Americas.)

MERLIN *Falco columbarius* 27–32cm, WS 64–70cm. Small dashing migrant falcon of varied open and semi-open habitats, from farmland and forest clearings to wetlands and cayes. Perches on fence posts, utility poles, in trees; rarely on wires, unlike American Kestrel. Wingbeats quick and powerful, hunting flight usually low and fast; soars on flattish wings, sometimes taking dragonflies in the air; does not hover. Generally rather dark overall; all plumages have blackish tail with narrow whitish to pale gray bars, overall dark underwings. Cf. Bat Falcon, much larger, relatively shorter-tailed Peregrine Falcon. Male slaty blue-gray above, female and imm. brownish above. **SOUNDS:** Mostly quiet; high screaming *kriih-kriih…* in interactions, with quality suggesting Killdeer. **STATUS:** Uncommon to fairly common Oct–Apr (a few from mid-Sep and into early May); most numerous as a transient in coastal areas. (Holarctic; winters in New World to S America.)

PEREGRINE FALCON *Falco peregrinus* 38–51cm, WS 96–119cm. Large, powerful migrant falcon of varied open and semi-open habitats, from wetlands and farmland to towns and cayes. Often perches on utility poles, towers in towns; not on wires. Soars on flattish or slightly raised wings; wingbeats smooth and powerful. Regularly hunts bats late in day, more so than Bat Falcon. Adult distinctive, with dark hood, barred underparts, slaty blue-gray upperparts (paler and bluer on male, darker and duskier on female); imm. has dark streaking below, cf. much smaller Merlin, which has relatively longer tail, narrower wings. Male obviously smaller than female when seen together. **SOUNDS:** Mainly in interactions, a harsh screaming *kaah-kaah…*, clipped clucking *kyuk-kyuk…*, and hoarse whistles. **STATUS:** Uncommon Oct–Mar; more widespread and locally fairly common on coast and cayes in migration, mid-Sep to Oct, Mar–Apr a few into May. (Almost worldwide.)

AMERICAN KESTREL

female

males

female

MERLIN

female

male

female

PEREGRINE FALCON

1st-year

adults

1st-year

TYPICAL OWLS (STRIGIDAE; 10+ SPECIES) Popular worldwide family of mainly nocturnal raptors, ranging from the tiny pygmy owls to the very large and spectacular Spectacled Owl. Ages differ, sexes similar but female larger; soft, downy-like juv. plumage worn briefly, soon replaced by adult-like feathers. In addition to songs, all owls have sundry other calls, mainly shrieks, mews, wails, and hisses given in various contexts, especially when breeding. Many species sing in duets, male song typically lower-pitched than female.

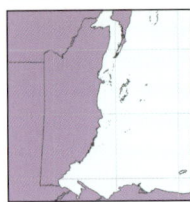

*NORTHERN MOTTLED OWL *Strix virgata* 33–38cm. Commonest and most familiar large owl in Belize; heard far more often than seen. Found in varied wooded and forested habitats, from rainforest edge to town parks, pine savanna, mangroves. Strictly nocturnal. Roosts at mid–upper levels in trees, typically in dense shady foliage; hunts from perches at low to mid-levels at edges of clearings, roadsides, other semi-open areas. Both morphs widespread; averages darker in wetter regions. Juv. has whitish facial disks, unstreaked buff underparts. **SOUNDS:** Song a measured series of deep, overslurred, slightly emphatic hoots, at a distance may suggest a dog barking; usually 3–6 notes (rarely to 10), often with louder notes toward the end before fading quietly with last 1–2 notes, *whuúh, whuúh, WHUÚH, WHUÚH, wuh*; typically about 1 note/sec, faster when excited. Fairly rapid bouncing-ball series of about 20 hoots suggests Spectacled Owl but usually longer, *wup wup wup-wup-wupwup…*, 5–6 notes/sec. Slurred wailing scream, about 1 sec, mainly by female. **STATUS:** Fairly common to common, including n. Ambergris Caye. (Mexico to nw. Peru.)

BLACK-AND-WHITE OWL *Strix nigrolineata* 38–41cm. Striking large owl of rainforest and edge, adjacent plantations. Strictly nocturnal. Mainly at mid–upper levels, often in canopy and subcanopy; sometime roosts in rather open situations. Visually distinctive, with bright yellow bill and feet, black face, barred whitish underparts. Cf. voice of Northern Mottled Owl. Juv. whitish overall with faint darker barring. **SOUNDS:** Typical song a slightly nasal, 2-note barking, 1st note quiet and inaudible at a distance, 2nd note loud, emphatic, *oh, WOAH!* every 15–30 secs. Longer series with quiet last note, *hoh-hoh-hoh-hoh-hoh, HWAOH, hoh*. Nasal quality distinct from gruffer, more resonant hoots of Northern Mottled Owl. Less often more measured short series hoots, *Hóah, Hóah,.....* Wailing scream lower, less shrieky than Northern Mottled. **STATUS:** Uncommon to scarce and perhaps local. (Mexico to nw. S America.)

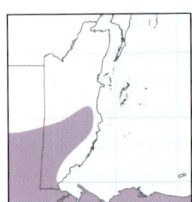

CRESTED OWL *Lophostrix cristata* 38–43cm. Spectacular large owl of rainforest, adjacent taller second growth. Strictly nocturnal. Calls mainly from canopy, where can be hard to see, but often roosts at low to mid-levels in shady understory. Dark and light morphs frequent, sometimes within pairs. Remarkable long bushy crests render ID straightforward, if you are lucky enough to see this stunning bird. Juv. whitish overall with dark facial disks, short whitish ear tufts, cf. juv. Spectacled Owl. **SOUNDS:** Song a deep, throaty, slightly overslurred growl, with short stuttering introduction audible at closer range, *k'k'k'Króhrrrr*, about 1 sec; repeated every 6–15 secs. **STATUS:** Uncommon to scarce, mainly in foothills of Maya Mts. (Mexico to S America.)

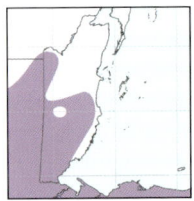

SPECTACLED OWL *Pulsatrix perspicillata* 43–48cm. Spectacular large owl of rainforest and edge, adjacent plantations. Nocturnal, but occasionally calls during day. Calls mainly from canopy, but roosts at any level, often in under-story whence may be flushed during day. Juv. whitish overall with dark facial disks, lacks whitish ear tufts of juv. Crested Owl. **SOUNDS:** Song a fairly rapid, pulsating series of usually 6–10 deep hoots, accelerating then fading, 1–2 secs; cadence suggests a sheet of metal being flexed quickly, *Wuup-wuup-wuupwuup…*; often given in overlapping duet of higher (female) and lower (male) songs. Cf. similar but usually longer variation of Northern Mottled Owl. Deep low *whoa*, suggesting large pigeon, sometimes given from roost on dull days. **STATUS:** Uncommon and local. (Mexico to S America.)

NORTHERN MOTTLED OWL

darker
morph

paler
morph

CRESTED OWL

BLACK-AND-WHITE OWL

paler
morph

barred
variant

darker
morph

SPECTACLED OWL

juv.

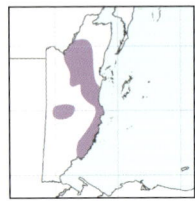

STYGIAN OWL *Asio stygius* 38–43cm. Large, dark horned owl of open pine forest, pine savanna; hunts at edges and in adjacent clearings. Roosts at mid–upper levels in larger trees, where can be conspicuous. Nocturnal; hunts from perches and in flight, often preying on bats. Distinctive, with overall dark plumage, whitish forehead patch, golden to amber eyes. Cf. Great Horned Owl. **SOUNDS:** Song a single low hoot, often with slightly muffled quality, *wuuh* or *woof*, typically every 3–10 secs; short, overslurred mewing scream, sometimes in duet with hoots. Snapping wing-clap in flight display. **STATUS:** Uncommon and local in Mountain Pine Ridge and lowland pine savannas. (Mexico and Caribbean to S America.)

STRIPED OWL *Asio clamator* 33–38cm. Pale horned owl of marshes, savanna, open country with scattered trees and thickets; roosts on ground and in dense trees. Nocturnal; hunts from perches such as fence posts, roadside wires. Flies with fairly shallow rapid wingbeats. Distinctive, with pale face, ear tufts, striped underparts, but cf. Barn Owl. **SOUNDS:** Song a low, moaning, over-slurred *hoóh*, every 4–18 secs. Piercing, hawk-like scream, *keeer*, about 0.5 sec, usually downslurred, sometimes in duet with hoots. Less often (agitated?) a fairly rapid series of about 12–20 short clucks or yelping hoots, about 3 notes/sec. **STATUS:** Uncommon and local in lowlands, mainly s. coastal belt; may be expanding range with deforestation. (Mexico to S America.)

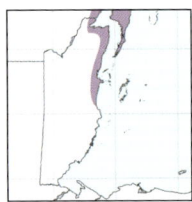

GREAT HORNED OWL *Bubo virginianus* 48–53cm, WS 105–124cm. Very large horned owl of semi-open habitats with scattered trees, forest patches, including towns and villages; not in humid forest. Roosts in trees, cliff cavities. Nocturnal, but may be seen at dusk and dawn, perched on roadside poles, wires, prominent snags. Hunts mainly from perches; flight strong and direct. Note very large size, rusty facial disks, white foreneck ruff, barred underparts. **SOUNDS:** Short series of deep resonant hoots, usually 5–8 notes with distinctive, slightly jerky cadence, a slower *whoo h'hoo, hoo, hoo*, and faster *whu h'h'hoo, hooo, hooo* often in duet. Juv. has raspy shriek, often repeated persistently, cf. Barn Owl. **STATUS:** Rare and local in n. coastal areas; very rare and sporadic s. of mapped range. (Americas.)

BARN OWLS (TYTONIDAE; 1 SPECIES) Worldwide family of owls with heart-shaped facial disks, relatively small eyes, short squared tails. Downy young molt direct into adult-like plumage.

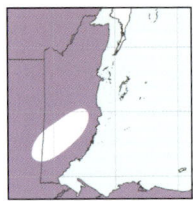

*BARN OWL *Tyto alba* 36–41cm, WS 93–110cm. Widespread and distinctive 'white owl' of open and semi-open country with scattered trees, forest patches, plantations, towns. Mostly nocturnal, but occasionally seen hunting late in day; often perches on roadside fence posts and wires. Roosts in caves, buildings, tree hollows, dense palm crowns. Hunts mainly in flight, quartering over open grassy areas with shallow easy wingbeats; often hovers. Underparts range from white to rich buff; male averages whiter below than female. **SOUNDS:** Far-carrying rasping shriek, often given in flight; other shrieks and hisses, none very heart-warming. **STATUS:** Uncommon to fairly common locally. (Worldwide.)

STYGIAN OWL

STRIPED OWL

GREAT HORNED OWL

BARN OWL

*FERRUGINOUS PYGMY OWL *Glaucidium brasilianum* 17–18cm. Small, relatively long-tailed owl active by day. Found in semi-deciduous and pine forest, gardens, plantations, mangroves, semi-open areas with hedgerows; avoids closed forest. Note voice, long tail (often twitched side-to-side), with numerous bars, diagnostic whitish streaking on crown. Juv. has plain crown or few pale markings. SOUNDS: Song an often prolonged series of hollow whistles, typically rather steady and often upslurred with 'inhaled' quality, *hoo hoo…*, or *whi' whi'…*; 10/3–3.5 secs. Bursts of high yelping twitters. STATUS: Fairly common to common in north and Mountain Pine Ridge, uncommon to rare and local in south. (Tropical Americas.)

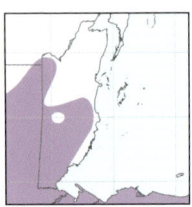

CENTRAL AMERICAN PYGMY OWL *Glaucidium griseiceps* 14–15cm. Tiny diurnal owl of rainforest and edge, mainly at mid–upper levels, often in subcanopy and usually not in open situations. Note spotted crown (plainer on juv.), relatively short tail with few pale bars; cf. Ferruginous Pygmy Owl. SOUNDS: Song a series of hollow ringing toots, usually starting with 2–4 notes, increasing to 6–9 notes (rarely up to 18); 10/2–3.5 secs. Song often preceded by soft quavering trills, also given alone and suggesting trill of Mesoamerican Screech Owl. STATUS: Uncommon. (Mexico to nw. Ecuador.)

*MESOAMERICAN [VERMICULATED] SCREECH OWL *Megascops guatemalae* 21–23cm. Small nocturnal owl of rainforest, overgrown plantations; mainly in forest interior and fairly dense vegetation where often difficult to see. Distinctive, the only screech owl in Belize. SOUNDS: Song an even-paced, rather soft purring trill, typically 4–15 secs, repeated after distinct pauses, usually of at least 10 secs. Easily confused with purring of Cane Toad *Rhinella marina*, especially at a distance; toad has throatier, more pulsating cadence suggesting a small motor. STATUS: Uncommon to fairly common. (Mexico to nw. Nicaragua.)

POTOOS (NYCTIBIIDAE; 2 SPECIES) Small New World family of large-headed nocturnal birds that pass the day roosting cryptically on branches and stumps. Ages differ slightly, with paler and looser-textured juv. plumage soon replaced by adult-like plumage; sexes similar. Readily detected at night by brilliant amber eyeshine reflected in light beams like burning coals; appreciably larger and brighter than nightjar eye reflection.

NORTHERN POTOO *Nyctibius jamaicensis* 38–43cm, WS 91–101cm. The common potoo in Belize; strictly nocturnal. Found in varied forested and semi-open habitats, even villages and towns. Roosts at mid–upper levels in trees, often in rather exposed situations. Hunts mostly from low to mid-level perches, including fence posts, less often high perches such as emergent snags, radio antennae. Hunting birds perch upright and sally for insect prey, often returning to same perch. Much larger than nightjars, and in flight as likely to be mistaken for a large hawk (much longer-tailed than owls); cf. rare Great Potoo. SOUNDS: Song (mainly spring–summer; at other seasons mostly on calm, often moonlit nights) a low, drawn-out rasp, about 1–1.5 secs, often followed by 1–6 gulping, upslurred clucks, *WAAAAHRR, wah wah wah*, repeated about every 6–20 secs. STATUS: Uncommon to fairly common locally. (Mexico and Caribbean to nw. Costa Rica.)

GREAT POTOO *Nyctibius grandis* 49–59cm, WS 112–125cm. Very large nocturnal bird that roosts and hunts from mid–upper levels in rainforest and edge, often in canopy or subcanopy along rivers. Like Northern Potoo, hunting birds perch upright and sally out for insect prey, often returning to same perch. Northern Potoo appreciably smaller with streaked vs. vermiculated plumage, dark mustache, golden-yellow eyes; but eyeshine of both species glowing amber at night. SOUNDS: Song 'unpleasant,' a throat-clearing, overslurred, moaning roar, *BWAAHr*, about 1 sec, occasionally followed by a quiet 2nd note; throatier and unhappier sounding than Northern Potoo, repeated about every 8–10 secs. In flight a shorter, more emphatic *woah!* STATUS: Scarce to uncommon and local; presence in Belize finally confirmed in late 2000s. (Mexico to S America.)

FERRUGINOUS PYGMY OWL

brown
morph

all pygmy owls have false
'eye-spots' on nape

rusty
morph

**CENTRAL AMERICAN
PYGMY OWL**

**MESOMERICAN
SCREECH OWL**

rusty
morph

brown
morph

NORTHERN POTOO

at night, potoo
eyes reflect like
burning coals

GREAT POTOO

roosting

NIGHTJARS (CAPRIMULGIDAE; 8 SPECIES) Worldwide family of nocturnal, insect-eating birds that sleep by day and rely on cryptic plumage to avoid detection and predators. Ages/sexes differ slightly in most species (mainly in tail pattern; females more often have rusty morph than males); like adult in 1 year. Most species nest on ground.

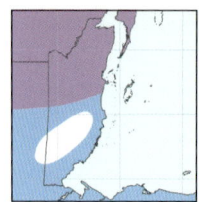

LESSER NIGHTHAWK *Chordeiles acutipennis* 20.5–23cm. Open and semi-open areas with scattered trees, forest edge, mangroves, towns; roosts on branches and ground, nests on ground. Mainly an aerial feeder, flight fairly erratic with bursts of quick flicking wingbeats and short glides on wings held in shallow V. Also hunts from ground, flushing off roads like typical nightjars. Main ID concern is Common Nighthawk; note that both species when perched usually show white on bend of wing, unlike other Belizean nightjars. Common has longer, more pointed wings, less fluttery flight with deeper, more rangy wingbeats; white wing band farther from wing-tip and slightly longer, extends across 5 primaries (4 primaries on Lesser); lacks bold pale barring across inner primaries. On perched Lesser, note grayish band of fine barring on breast (lacking on Common); white wing band of male lies under tip of tertials, vs. forward of tertials on Common; female has buff wing band. On both species, wing-tips can project past tail at rest, more often on Common. **SOUNDS:** Song (spring–summer) from ground or low perch a churring trill, swells quickly and fades abruptly, *urrrr…*, can go on for minutes, typically in bursts of 7–20 or so secs with pauses of 1–3 secs; can be mistaken for Cane Toad *Rhinella marina*, which is lower pitched, more pulsating. Bleating *whik* in flight, mainly in interactions. Flushed winter birds usually silent. **STATUS:** Uncommon to fairly common Sep–Apr; local breeder in north, including larger cayes. (Mexico and sw. US to S America.)

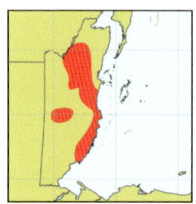

COMMON NIGHTHAWK *Chordeiles minor* 23–24cm. Breeds in pine savanna and Mountain Pine Ridge; transient migrants may occur in any habitat; roosts mainly on branches, nests on ground. Habits like Lesser Nighthawk but flight stronger, less fluttery; in breeding season often flies much higher. Undergoes wing molt in S America, whereas Lesser molts late summer–fall in Belize. Plumage tones more variable than Lesser; migrants vary from dark overall to relatively pale and sandy gray. Cf. Lesser Nighthawk. **SOUNDS:** Sharp nasal *beenk* or *peehn* in flight; in display, male stoops steeply and produces rushing boom at bottom of dive. **STATUS:** Uncommon to fairly common locally Apr–Sep, mainly in north; more widespread and sporadically common in migration, mainly late Aug–Oct, Apr–May; rarely into Nov and from mid-Mar. (Breeds N America to Panama, winters S America.)

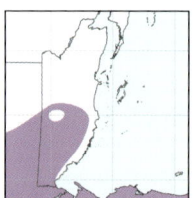

SHORT-TAILED NIGHTHAWK *Lurocalis semitorquatus* 20–21cm. Poorly known in Belize. Inhabits rainforest and edge, especially along rivers and streams; roosts and nests in forest canopy. Singles or pairs emerge at dusk to feed low over canopy; erratic flight suggests a large bat rather than *Chordeiles* nighthawks. Distinctive, with fairly large size, short tail, overall dark plumage (no white wing bands). Sexes similar. **SOUNDS:** Sharp, slightly liquid clucks, singly or in short quick series, *g'wik* and *gwik whik-whik*. **STATUS:** Scarce and local (overlooked?) in south. (Mexico to S America.)

(COMMON) PAURAQUE *Nyctidromus albicollis* 26–30cm. Large familiar nightjar of roadsides, humid forest edge, pastures, mangroves. Nests and mainly roosts on ground, in leaf litter under bushes and trees. Hunts from ground and low perches, sallying out with stiff, flicking wingbeats and flat-winged glides, often returning to or near the same spot. Note very long tail at rest, cf. *Antrostomus* nightjars; also bold scapular pattern, rows of buff spots on wing coverts, pale underparts with fine dark barring. White wing and tail flashes of male striking and distinctive; note buff wing band of female, paler wing band. imm. male resembles female but with more white in tail. **SOUNDS:** Song (mainly spring–summer) a loud, slightly burry whistled *pWEER!* every 2–3 secs; year-round a quieter *p'weéir*, often preceded by hesitant, stuttering clucks. Nervous quiet clucks from perched birds. **STATUS:** Fairly common to common. (Mexico and s. Texas to S America.)

LESSER NIGHTHAWK

female

male

breeder

COMMON NIGHTHAWK

males

male

transient

female

SHORT-TAILED
NIGHTHAWK

female

PAURAQUE

males

rusty morph

YUCATAN POORWILL *Nyctiphrynus yucatanicus* 20.5–21.5cm. Small dark nightjar of humid semi-deciduous forest and edge, adjacent second-growth thickets. Hunts mainly from mid–upper levels in trees, less often from ground; often sings from perch in dense cover. Note voice, rather plain face and breast with narrow white forecollar, white dots on upperwing coverts, small white tail corners. Cf. appreciably larger Yucatan Nightjar, which overlaps widely. Sexes similar. **SOUNDS:** Song persistent in spring–summer, sometimes for brief periods around dusk and dawn at other seasons, a resonant, slightly burry *wheeéu*, steadily every 1–2 secs; cf. Pauraque. Clucking *puk*, singly or in short series, when agitated. **STATUS:** Uncommon to locally fairly common. (Mexico to n. Belize.)

YUCATAN NIGHTJAR *Antrostomus badius* 24–25.5cm. Rather large nightjar of humid forest and edge, clearings, scrub, mangroves. Sings and hunts mainly from perches; feeding birds often at mid-levels along edges, sometimes on fence posts. Note voice, relatively dark plumage with tawny-cinnamon hindcollar, coarse white spotting on underparts, tail pattern (especially extensive white of male, and strong graduation). Cf. other nightjars. **SOUNDS:** Song persistent in spring–summer, sometimes for brief periods around dusk and dawn at other seasons, a loud whistled *tu weéu-weéu*, every 1–2 secs, may suggest Chuck-will's-widow; 1st note quiet, inaudible at a distance. Call a hard, hollow clucking *k-lok*, often in short series. **STATUS:** Uncommon to fairly common mid-Oct to Apr; breeds locally in north but status in many areas awaits elucidation; scarce transient on cayes. (Mexico to Belize, winters to n. Honduras.)

EASTERN WHIP-POOR-WILL *Antrostomus vociferus* 23–24cm. Poorly known in Belize. Medium-size, nonbr. migrant found in forest and edge, second growth, scrub. Note relatively gray plumage with slightly paler and grayer crown sides and scapulars, narrow and usually indistinct buff hindcollar, tail pattern; cf. Chuck-will's-widow, Yucatan Nightjar. **SOUNDS:** Rarely sings in Belize, but possible in spring migration, a whistled *whie-pura-weén*, every 1–1.2 secs. Flushed birds may give hollow clucks. **STATUS:** Scarce or uncommon nonbr. migrant; most records Oct–early Nov and Mar, but likely overlooked in winter, when might be found anywhere. (Breeds e. N America, winters s. to Costa Rica.)

CHUCK-WILL'S-WIDOW *Antrostomus carolinensis* 29–33cm. Large nonbr. migrant nightjar found in humid forest and edge, second growth, scrub, plantations, mangroves; migrants regular on cayes. Roosts and hunts from ground and perches, sometimes at mid–upper levels in trees. Note large size and very long wings, overall rich dark brown to gray-brown plumage without striking contrast except male tail pattern (from above shows only on spread tail); tail rusty brown above vs. grayish on Eastern Whip-poor-will. Cf. other nightjars. **SOUNDS:** Migrants sometimes sing briefly in spring, usually for short periods around dawn and dusk, a rapid *chk weéu-weéu*, intro cluck inaudible at a distance; cf. song of Yucatan Nightjar. Typically silent when flushed. **STATUS:** Scarce to uncommon, especially in fall on cayes; most records Sep–Oct and Mar–early Apr, but likely overlooked in winter, when might be found anywhere. (Breeds e. US, winters to n. S America.)

YUCATAN POORWILL

rusty morph

YUCATAN NIGHTJAR

1st-year
female

adult
female

male

EASTERN WHIP-POOR-WILL

female

male

male

CHUCK-WILL'S-WIDOW

female

male

female

PIGEONS (COLUMBIDAE; 19 SPECIES) Familiar, worldwide family of rather plump-bodied birds with small heads, relatively short sturdy legs. Ages differ slightly, sexes similar (female often averages duller than male) or strikingly different; adult appearance attained in 1st year.

MOURNING DOVE *Zenaida macroura* 25.5–31cm. Scarce nonbr. migrant to open and semi-open areas, especially farmland, coastal scrub. Feeds on ground, where easily overlooked until flushed; flight fast and direct, usually fairly low. Note long pointed tail (longer on male), big dark spots on wing coverts. Female duller than male, without blue-gray hindneck. **SOUNDS:** Rarely vocal in Belize, but in flight and when flushed makes high rapid wing whistle. **STATUS:** Scarce to locally uncommon Oct–Apr, mainly in north; most frequent in fall migration, generally rare and local in winter. (N America to Panama.)

WHITE-WINGED DOVE *Zenaida asiatica* 27–30.5cm. Attractive dove of semi-open areas, especially ranchland, towns, villages, brushy woodland edge. Feeds mainly on ground but often perches on roadside wires. Flight often higher overhead than Mourning Dove, more like Red-billed Pigeon and Eurasian Collared Dove. Note bold white wing band, white tail corners; cf. larger and paler Eurasian Collared Dove. **SOUNDS:** Short song a mournful, slightly hoarse cooing *wh-koó ku-kooo* ('who cooks for you'); long song lower, slightly chanting, a 3-note then 4-note phrase and varied ending of 1–4 notes, such as *h-hoo-coo, h-hoo coo-oo, oo oóo oo*. **STATUS:** Common in north, including larger cayes; more widespread in migration and winter; fairly common Aug–Nov in south, where scarce to uncommon and local at other seasons. Increasing with deforestation; first recorded Belize in late 1960s. (Mexico and sw. US to Panama.)

EURASIAN COLLARED DOVE *Streptopelia decaocto* 31–34cm. Non-native dove increasing in open and semi-open areas, mainly in towns and villages, ranchland. Often on roadside wires and utility poles, locally in flocks. Note plain upperparts with narrow black hindcollar, big white tail corners; plumage tones variable, some pale and milky, others rather dark. **SOUNDS:** Song a mournful 3-syllable cooing *wh'Huuu hu*, about 1 sec, often repeated several times in tedious succession. Flight call a burry or snarling *réhhr*. **STATUS:** Fairly common locally, especially in north, but increasing. First recorded Belize in mid-1990s, established only in 2010s. (Native to Eurasia.)

FERAL PIGEON (ROCK DOVE) *Columba 'livia'* 30.5–35cm. Widespread, widely bastardized human commensal of towns, urban areas; rarely far from habitation; in some areas may revert to nesting in cliffs, its native habitat. Often in flocks on roadside wires, buildings; infrequently perches in trees. Plumage highly variable, an annoying source of potential confusion with native pigeons, but note habitat and habits; in flight, underwing coverts often white (dark on native pigeons). **SOUNDS:** Low muffled cooing audible at close range. **STATUS:** Fairly common in most urban areas; scarce and local in villages away from main highways. (Native to Eurasia.)

WHITE-CROWNED PIGEON *Patagioenas leucocephala* 33–37cm. Large dark pigeon of mangroves, semi-deciduous forest and edge. Mainly at mid–upper levels in canopy, sometimes on bare snags. Like other large native pigeons, usually wary and mostly seen in flight. Note solidly dark plumage, pale yellowish eye; snow-white cap of adult. Cf. Red-billed and Feral Pigeons. **SOUNDS:** Low, slightly burry cooing song typically a short note followed by a repeated 4-syllable phrase (usually 2–5×), *whoo, whoo-t-huh-whooor, whoo-t-huh-whooor....* Paired rising and falling purring growls usually preceded by a quiet cluck, *uk purrrrr urrrr*. **STATUS:** Locally/seasonally uncommon to fairly common on cayes, mainly late Feb–Oct, when scarce and irregular on mainland; scarce on cayes in winter, when more frequent in mainland coastal areas; very rare and sporadic well inland in north, mainly Dec–Apr. (Caribbean region.)

MOURNING DOVE

male

WHITE-WINGED DOVE

EURASIAN
COLLARED DOVE

FERAL PIGEON

WHITE-CROWNED
PIGEON

juv.

PALE-VENTED PIGEON *Patagioenas cayennensis* 30–34cm. The common large gray pigeon of savanna, gallery forest, humid forest edge, towns and villages in coastal lowlands; not in heavy rainforest. Mainly at mid–upper levels, often perched in canopy, on roadside wires. In display flight, climbs with exaggerated slow wingbeats, glides down with wings in V. Note blue-gray head with red eye, dark bill; whitish belly contrasting with grayish tail. Cf. Red-billed and Scaled Pigeons. Juv. duller overall with narrow pale edgings to upperparts. **SOUNDS:** Song a drawn-out coo followed by a repeated 3-syllable phrase (usually 3–4x): *whoooo, oo-k-hoooo, oo-k-hoooo…*, cf. Red-billed Pigeon song; deep purring *whoor*. **STATUS:** Fairly common to common, especially in coastal belt and south; occasional visitor to inshore cayes. (Mexico to S America.)

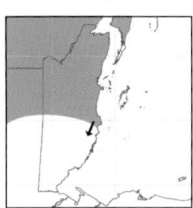

RED-BILLED PIGEON *Patagioenas flavirostris* 31–35.5cm. Large dark pigeon of semi-deciduous forest and edge, gallery forest, mangroves; not in rainforest. Mainly at mid–upper levels, often perched on bare snags in canopy and clearings. Often looks simply dark overall; 'red' bill is mostly pale yellowish with small reddish area at base often hard to see. Note pinkish head and neck, slaty blue-gray belly contrasting with black tail. Cf. Pale-vented, Short-billed, and Scaled Pigeons. Juv. has rustier tone to chest and scapulars. **SOUNDS:** Song a drawn-out coo followed by a repeated 4-syllable phrase (usually 2–5×): *whoooooo, óo k-hoo-oo, óo k-hoo-oo…*, cf. Pale-vented Pigeon song; swelling *whoo*, often repeated several times. **STATUS:** Fairly common and increasing in north; scarce and sporadic but increasing in south. (Mexico to Costa Rica.)

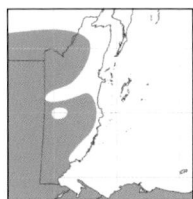

SHORT-BILLED PIGEON *Patagioenas nigrirostris* 27.5–29cm. Relatively small dark pigeon of rainforest and edge. Often heard but infrequently seen. Mainly at mid–upper levels in canopy; most often seen as singles or pairs in flight over canopy or across small openings, not in flocks; sometimes found at fruiting trees, or along quiet roadsides taking grit. Dark overall and rather nondescript; note small black bill, brownish upperparts, voice. **SOUNDS:** Song a far-carrying, relatively high 4-syllable cooing, *ooh' HOO-ku-koo* with slightly jerky cadence, repeated every few secs; lacks introductory notes of larger pigeons; deep purring *urrrrr*, at times repeated steadily. **STATUS:** Fairly common, especially in south. (Mexico to nw. Colombia.)

SCALED PIGEON *Patagioenas speciosa* 30.5–35cm. Large handsome pigeon of humid forest and edge. Mainly at mid–upper levels in canopy; overlooked easily but sometimes perches on bare snags in canopy and clearings. Mostly seen in flight as singles or small groups, but locally large flocks may gather at seasonal food sources. Note white vent and undertail coverts contrasting with black tail; flight slightly 'looser' and floppier than other pigeons; boldly scaled neck and breast distinctive when visible. Cf. Red-billed and Pale-vented Pigeons. Juv. duller overall with indistinct scaling. **SOUNDS:** Song a deep, slightly moaning single note followed by a repeated 2–3-syllable phrase (usually 2–4×): *whoooo huh-woóhr, huh-woóhr…* and variations; overall more sonorous, slower-paced than other lowland forest pigeons. **STATUS:** Uncommon to fairly common but local and nomadic; seasonal movements poorly known. (Mexico to S America.)

PALE-VENTED
PIGEON

RED-BILLED
PIGEON

SHORT-BILLED
PIGEON

SCALED PIGEON

male

female

GENUS *COLUMBINA* (4 species). Very small, easily overlooked ground doves of open habitats, often along roadsides; perch readily on fences and in low trees. Flush explosively from ground with variable wing whirr; all species have rusty wing flashes. Frequently form flocks, and species mix readily.

COMMON GROUND DOVE *Columbina passerina* 16–17cm. Very small dove of open and semi-open areas, from pine savanna and coastal scrub to towns to farmland. Feeds on ground, mainly as pairs or small groups. Note scaly neck and breast (old name of Scaly-breasted Ground Dove much more appropriate), orange to pinkish bill base; dark marks on upperwing coverts more extensive than Ruddy and Plain-breasted Ground Doves. In flight, paler and grayer overall than Ruddy, with dark secondaries, pale upperwing band; lacks black axillars. **SOUNDS:** Song a low hooting *whuuh'* or slightly disyllabic *h'wooh*, repeated steadily every 1–1.5 secs. Flushes with quiet wing whirr. **STATUS:** Fairly common in n. coastal belt, including Ambergris Caye; sporadic but spreading farther inland and south, mainly fall–winter; ranges to inshore cayes. (Mexico and s. US to S America.)

PLAIN-BREASTED GROUND DOVE *Columbina minuta* 14.5–15.5cm. Very small dove of pine savanna, also locally in farmland with scattered bushes and trees. Feeds on ground, mainly as pairs or small groups; sometimes mixes with other ground doves. Mostly inconspicuous and overlooked easily, but sings from low perch such as fence or shrub. Note small size, plain neck and breast, purple sheen to dark upperpart markings; bill base can be dull pinkish, cf. Common Ground Dove. Appreciably smaller, shorter-tailed, and grayer overall than female Ruddy Ground Dove, with dark secondaries, lacks black axillars. **SOUNDS:** Song a low hooting *wüp* or *w'up* repeated steadily every 0.5–1 secs; notes much shorter and song faster-paced than Common Ground Dove. **STATUS:** Uncommon to fairly common but rather local; spreading with deforestation. (Mexico to S America.)

RUDDY GROUND DOVE *Columbina talpacoti* 17–18cm. Common small dove, widespread in open and semi-open areas. Feeds on ground, locally in flocks up to 100 or so birds. Male distinctive; female best told from other *Columbina* ground doves by relatively large size, plain neck and breast, warm plumage tones. Also cf. female Blue Ground Dove. **SOUNDS:** Song a 2-syllable hooting *per-woop* or *h'woop*, repeated steadily every 0.5–1 secs, quieter 1st note not always audible at a distance, when cf. Plain-breasted Ground Dove; at times varied to 3-syllable *h't'woop*. **STATUS:** Common to fairly common virtually throughout, but rare at higher elevations, including Mountain Pine Ridge; very rare and sporadic on inshore cayes. (Mexico to S America.)

BLUE GROUND DOVE *Claravis pretiosa* 20–21.5cm. Attractive small dove of humid forest edge and clearings, second growth. Most often seen in flight. Feeds on ground as singles, pairs, small groups; sings from mid–upper levels in trees, often well concealed. Powder-blue male distinctive, even with a brief flight view. Female has contrasting rusty rump and tail, broad chestnut wingbars, grayish underwings, cf. smaller Ruddy Ground Dove. **SOUNDS:** Song a far-carrying, clear hooting *booop* or *oop* every 1.5–2 secs; usually 2–10× in unhurried, measured series; typically 3–15 sec pauses between series. **STATUS:** Uncommon to fairly common but local, most numerous in south; possible local movements await study. (Mexico to S America.)

INCA DOVE *Columbina inca* 20–22cm. Rare but increasing. Long-tailed ground dove of open and semi-open areas, roadsides, towns. Singles or pairs associate readily with Common and Ruddy Ground Doves; often perches on roadside wires. Distinctive, with scaly plumage, long squared tail with white sides; bright rusty wing patches. **SOUNDS:** Song a pair of hollow, slightly burry, overslurred coos, *whooh pooh*, every 1–2 secs, often repeated steadily; less often a faster, burry *h'p'wuurrr....* Wings make whirring rattle when flushed. **STATUS:** Rare (mainly Aug–Nov, Feb–Jun) and sporadic, spreading with deforestation from Guatemala; first recorded Belize 1996, most records since mid-2010s. (Mexico and sw. US to Costa Rica.)

COMMON GROUND DOVE

female

male

PLAIN-BREASTED GROUND DOVE

female

male

RUDDY GROUND DOVE

female

male

BLUE GROUND DOVE

female

male

INCA DOVE

GENUS *LEPTOTILA* (4 species). Rather plain, fairly plump doves of forest floor; mostly detected by voice. Typical view is a bird flushing off into forest understory or flying low and fast across a road or trail. All species have white tail corners ('white tips'). Often best seen early and late in day on quiet roadsides and at edges. Sometimes land on low branch after being flushed. Sing from ground or low perch.

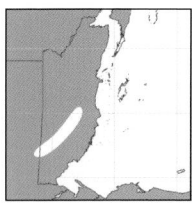

*LAWRENCE'S [WHITE-TIPPED] DOVE *Leptotila [verrauxi] fulviventris* 26–29cm. Widespread *Leptotila*, found in a variety of forest and edge habitats; absent at higher elevations and from s. rainforest. See genus intro. Note voice, overall rather plain pinkish head and breast, vs. stronger head and face patterns of other *Leptotila*. Juv. duller overall, upperparts and breast with pale edgings. SOUNDS: Song a low, mournful, 2–3-syllable cooing, *ooh-wooooo*, 2nd note sometimes inflected, or *ooh'h-woooo*, every 3–10 secs; 1st note may be inaudible at a distance Sometimes preceded by 1 or more low purring *krrrr* notes. STATUS: Fairly common to common, especially in north. (Mexico and s. Texas to Nicaragua.)

*YUCATAN [CARIBBEAN] DOVE *Leptotila [jamaicensis] gaumeri* 26–28cm. Local in n. Belize, in semi-deciduous forest, wooded thickets. See genus intro. Note voice, contrasting white face and blue-gray crown; cf. Lawrence's Dove (occurs in same areas), smaller and duskier Gray-headed Dove. White tail tip more extensive than other *Leptotila*, across 5 feathers (vs. 3–4 on Lawrence's, 3 on Gray-headed), and underwings more extensively bright rusty than Lawrence's. SOUNDS: Song a low, mournful 4-syllable cooing, *wh-hoo-oo-ooo* or *hu cooo'hu-ooo*, last note fading away; every 7–12 secs. STATUS: Fairly common in n. Corozal, including Ambergris Caye, scarce and local farther south. (Mexico to n. Honduras.)

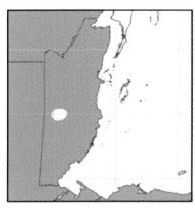

GRAY-HEADED DOVE *Leptotila plumbeiceps* 25.5–27.5cm. Widespread in humid forest and edge, adjacent plantations, second growth. See genus intro. Note voice, blue-gray crown and nape, dusky pinkish breast; white tail corners smaller than Lawrence's and Caribbean Doves. Gray-chested Dove darker overall with warm brown crown and nape. Juv. duller overall, upperparts edged cinnamon, breast narrowly scalloped buff. SOUNDS: Short, mournful, slightly overslurred cooing *whooo* or *huuu*, about 0.5 sec, repeated steadily every 2–3 secs, cf. lower, more fading away, and often longer coo of Ruddy Quail-Dove. STATUS: Fairly common in south, uncommon and more local in north. (Mexico to w. Colombia.)

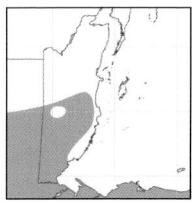

GRAY-CHESTED DOVE *Leptotila cassinii* 25–27cm. Found in rainforest and edge, adjacent plantations, second growth, *Heliconia* thickets. See genus intro. Distinctive, rather small and dark *Leptotila* with warm brown crown and nape, dusky grayish breast, small white tail corners. Cf. Lawrence's and Gray-headed Doves, the 3 species potentially audible alongside each other in some places. Also cf. song of Ruddy Quail-Dove. Juv. duller and darker overall with variable cinnamon feather edgings. SOUNDS: Song a low, mournful, drawn-out, fading away *whoooo* or *whoo6oo*, about 1.5 secs, every 4.5–6.5 secs; cf. shorter, faster-paced song of Ruddy Quail-Dove. STATUS: Fairly common. (Mexico to nw. Colombia.)

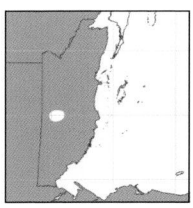

RUDDY QUAIL-DOVE *Geotrygon montana* 23–25cm. Elusive, rather chunky terrestrial dove of humid forest, shady plantations; mostly detected by voice. Singles or pairs walk on forest floor, running quickly when alarmed; also may freeze and flush explosively from close range with wing whirr, like quail. Sing from ground and low to mid-level perch. Bright ruddy male readily identified; female slightly smaller, stockier, and shorter-tailed than *Leptotila* doves, with no white on tail corners; also note striped face, vertical pale bar at breast sides, red bill. SOUNDS: Song a low, mournful, slightly moaning *whoooo*, fading away slightly, every 2–5 secs; cf. song of Gray-chested Dove. STATUS: Uncommon to fairly common on mainland; very rare wanderer to cayes. (Mexico to S America.)

LAWRENCE'S DOVE

YUCATAN DOVE

GRAY-HEADED DOVE

juv.

GRAY-CHESTED
DOVE

RUDDY QUAIL-DOVE

juv.

female

male

NEW WORLD PARROTS (PSITTACIDAE; 9+ SPECIES) Familiar group of brightly colored, often noisy birds associated with the tropics. Ages/sexes similar or slightly different; mostly attain adult appearance in 1st year. Several species reduced in numbers by trapping for pet trade. Often seen in flight, and getting good views of perched birds can be challenging.

AMAZONS (GENUS *AMAZONA*)

(5+ species). Medium-size to large 'typical' parrots with rounded wings, relatively short squared tails, red upperwing patches, and species-specific head patterns. Fly with distinctively shallow, fairly rapid wingbeats; pairs often keep together within larger groups. Ages/sexes differ slightly or similar. Vocalizations notably varied, but with experience the quality of sounds and some key phrases can be helpful in ID.

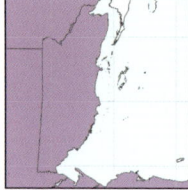

RED-LORED AMAZON (PARROT) *Amazona autumnalis* 32–35.5cm. Commonest and most conspicuous large amazon in much of Belize, found in humid forest and edge, semi-open areas with forest patches, taller trees. In pairs and loose flocks; typically flies high overhead, feeds mainly in canopy. Note contrasting yellow cheeks (reduced or lacking on imm.), dusky gray bill, voice. Cf. slightly larger and bulkier Northern Mealy and Yellow-headed Amazons. SOUNDS: Varied raucous screaming, often with fairly shrill, yapping quality, *rriek-rriek rriek-i-rrak*, and *zi-kreek*; quieter and mellower calls mainly when perched. STATUS: Fairly common to common almost throughout, becoming uncommon and local in Corozal. (Mexico to nw. S America.)

*NORTHERN MEALY AMAZON (PARROT) *Amazona [farinosa] guatemalae* 38–43cm. Large amazon of rainforest and edge, not in open or semi-open country with scattered trees and forest patches. In pairs and loose flocks; heard more often than seen in its forested habitat, where feeds mainly in canopy. Distinctive, with broad pale eyering often more noticeable than blue crown; hindneck often has pale, 'mealy' scalloping. Juv. has duller bluish crown. Cf. other amazons, which favor edge and more open habitats. SOUNDS: Loud and raucous, but relatively mellow and low-pitched, including *churruk-churruk, chuk chuk cheeurr*; a deep slightly gruff *rrehk-rrehk…*, and *rrah krrah-krrah*. Often silent in flight. STATUS: Uncommon to locally common, especially in south. (Mexico to w. Panama.)

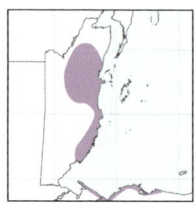

*YELLOW-HEADED AMAZON (PARROT) *Amazona [ochrocephala] oratrix* 35.5–38cm. Declining large amazon of pine savanna, gallery forest, adjacent semi-open country with forest patches, taller trees. In pairs and loose flocks, seen mainly early to mid-morning and late in the day commuting to and from feeding areas. Often detected by voice, which carries well. Distinctive if seen, with bright yellow face and crown, pale bill; also note small red leading edge to wing. Juv. has duskier bill, yellow limited to crown and lores; probably develops full yellow head in 1–2 years. Overlaps widely with much commoner Red-lored Amazon, and may associate loosely with that species. SOUNDS: Raucous but relatively deep and mellow calls have vaguely human quality, including rolled *kyaa'aa'aah* and *krra'aah'aow*, deep rolled *ahrrr*, throaty *rrohrr*, etc. Often silent in flight. STATUS: Uncommon to fairly common but local; widely extirpated by capture for pet trade; also persecuted locally by citrus farmers. (Mexico to Honduras.)

pairs of amazons tend to keep
together within flocks

RED-LORED AMAZON

imm.

NORTHERN
MEALY AMAZON

YELLOW-HEADED AMAZON

imm.

WHITE-FRONTED AMAZON (PARROT) *Amazona albifrons* 25.5–27cm. Small amazon of semi-deciduous forest and edge, semi-open areas with scattered trees and forest patches, mangroves, locally in towns. In pairs and loose flocks; often flies lower than large amazons, at or below canopy height, but roost flights can be high overhead. Wingbeats relatively quick, flight looks hurried relative to large amazons. Yucatan Amazon in flight typically has paler green rump and tail, less red on male's primary coverts (but female often has some red, unlike female White-fronted); also note Yucatan's dark cheek patch, stronger dark scalloping on head and body; yellow lores visible with good views. **SOUNDS:** Raucous screaming and yapping, often faster-paced than large amazons, including sharp *kyi kyeh-kyeh…*, and rapid, often paired yapping *kyak-yak-yak-yak, yak-yak*. Mellower rolled and screeching calls especially when perched. Cf. Yucatan Amazon. **STATUS:** Fairly common to common. (Mexico to Costa Rica.)

YUCATAN (YELLOW-LORED) AMAZON (PARROT) *Amazona xantholora* 25.5–28cm. Small amazon of semi-deciduous forest and edge, semi-open areas with scattered trees and forest patches, coastal scrub, mangroves. Habits much like White-fronted Amazon, with which it overlaps widely; the 2 species occur alongside each other but appear not to mix in flocks; unclear what separates them ecologically. Best distinguished from White-fronted by head pattern, upperwing pattern in flight (see that species for details); some females lack red on primary coverts, like White-fronted. Imm. male resembles female but averages more red on primary coverts. **SOUNDS:** Raucous and varied, similar to White-fronted Amazon but may average mellower, including rolled *reeéeah*, and *kyeh-kyeh keeei'iirr*; yapping calls average lower, burrier, *rek-rek-rek-rek rek-rek rek-rek….* **STATUS:** Fairly common to common in Corozal, becoming uncommon and more local to the south. (Mexico to n. Honduras.)

WHITE-CROWNED PARROT *Pionus senilis* 23–25.5cm. Medium-size, rather compact parrot of humid forest, pine forest, adjacent semi-open areas with forest patches. In pairs or small flocks, at times of 50 or so birds. Flies with distinctive deep wingbeats, unlike shallow flapping of amazons and hurried flight of Brown-hooded Parrot. Feeds mainly in canopy, at times perching on exposed snags. Distinctive, with extensively blue plumage, big white crown patch, mostly red undertail coverts but no red in wings. **SOUNDS:** Raucous screeches in flight, *rreéahk* and rolled *rriéah*, etc, higher and screechier than amazons, deeper and less metallic than Brown-hooded. **STATUS:** Fairly common to common. (Mexico to w. Panama.)

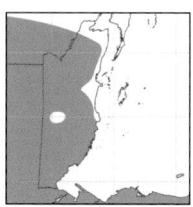

BROWN-HOODED PARROT *Pyrilia haematotis* 20.5–23cm. Small, rather compact parrot of rainforest and edge. In pairs or small groups, typically seen briefly in flight at canopy level. Flight fast and direct with quick deep wingbeats, always seems in a hurry compared with slightly larger White-crowned Parrot, which has slower deep wingbeats, often flies higher. Feeds at mid–upper levels in fruiting trees, where quiet and overlooked easily. Note voice, dark brown head with pale spectacles, bright red 'armpit' flashes in flight. Juv. has paler brown hood, duller ear spot, greener chest. **SOUNDS:** Shrill, slightly metallic screeches in flight, *kreéik* or *kreeíh'* and short series, *kreiik krríik*; higher, more metallic than White-crowned Parrot. **STATUS:** Fairly common to common in south and west, uncommon and more local in northeast. (Mexico to nw. Colombia.)

*AZTEC [OLIVE-THROATED] PARAKEET *Eupsittula [nana] astec* 23–25cm. The only parakeet in Belize. Favors humid forest and edge, semi-open areas with trees, hedgerows, second growth, mangroves. In pairs or small flocks, locally to 30 or so birds; flight fast and direct to twisting, mainly at around treetop height. No similar species in Belize; note long pointed tail, plain head, blue in wings. **SOUNDS:** Screechy, slightly burry *krrieh* and *kreeíh*, often doubled and in screaming series; also lower burry calls when perched, *rreh* and *krreh*. **STATUS:** Fairly common to common; occasional visitor to Caye Caulker in winter. (Mexico to w. Panama.)

WHITE-FRONTED AMAZON

imm.

females

males

YUCATAN AMAZON

female

male

female

male

WHITE-CROWNED PARROT

juv.

BROWN-HOODED PARROT

AZTEC PARAKEET

adult variation

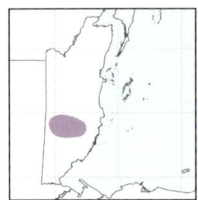

SCARLET MACAW *Ara macao* 81–96cm. Spectacular, large and very long-tailed parrot of rainforest and edge, tree-scattered pastureland with forest patches, often near rivers; breeds in tree cavities. No similar species in Belize. In pairs or small groups, ranging widely for food; mainly seen in flight early to mid-morning and late in the day commuting to and from feeding areas or near roosting and nesting sites. At other times can sit quietly in canopy, where easily overlooked. Flight direct and relatively unhurried, with strong steady wingbeats, tail flowing out behind. SOUNDS: Far-carrying, deep raucous cries in flight, *rrah*, and *rrrahk*; quieter calls sometimes when perched. STATUS: Scarce, declining, and local; Belize population perhaps only around 150 birds. Core range is in se. Cayo, with regular seasonal movement (mainly Jan–Mar) to areas e. of Maya Mts. (Mexico to S America.)

CUCKOOS (CUCULIDAE; 8+ SPECIES) Worldwide, notably diverse family found mainly in warmer climates; most species rather long-tailed, all have 2 toes pointing forward, 2 backward. Ages similar or different, attaining adult appearance in 1st year; sexes similar.

*COMMON SQUIRREL CUCKOO *Piaya cayana* 40–47cm. Spectacular and familiar tropical cuckoo of forest, woodland, and edge, semi-open areas with hedgerows, scrubby thickets with taller trees, mangroves. Mainly at mid–upper levels, where can be surprisingly difficult to see; hops and runs along branches and amid foliage a little like a squirrel; typically sings from concealed perch. Often seen flying across roads; flights usually short, bursts of wingbeats alternated with sweeping glides. No similar species in Belize. SOUNDS: Sharp, woodpecker-like *chik!* sometimes doubled, often followed by overslurred low scream, *chik! reowh*; gruff, slightly raspy stuttering *ehk'ehr-rrer*. Song (mainly spring–summer) a usually fairly steady-paced, often prolonged series of overslurred sharp whistles, *wheep wheep…*, typically 10 notes/5.5–8 secs. STATUS: Fairly common. (Mexico to S America.)

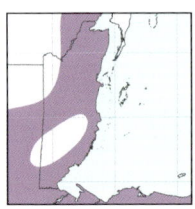

STRIPED CUCKOO *Tapera naevia* 28–30.5cm. Open and semi-open areas with brushy thickets, second growth, overgrown grassy fields, forest edge. Rarely seen unless singing (thus appears rare or absent from many areas in winter), when often perches on fence post, roadside wires; otherwise remains on or near ground in cover. Brood parasite, typically of species with domed or covered nests. Distinctive, with striped upperparts, erective spiky crest (raised and lowered when singing), plain pale underparts; black 'wrist flags' rarely visible, flashed in display. SOUNDS: Song (spring–summer) far-carrying, a clear, deliberate double whistle, easily imitated, *wheee whee*, 2nd note slightly higher; less often longer series, *whee' whee' whee' whee'buh*, and variations. STATUS: Uncommon to fairly common. (Mexico to S America.)

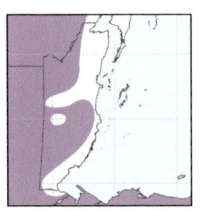

PHEASANT CUCKOO *Dromococcyx phasianellus* 35.5–37cm. Scarce cuckoo of forest and edge, adjacent taller second growth. Rarely seen unless singing (thus appears rare or absent from many areas in winter), when often perches high in trees; otherwise remains on or near ground, where elusive. Flight direct, wingbeats mostly below body plane. Brood parasite, typically of species with domed or pendulous nests. Distinctive, with small head, slender bill, pointed crest, and long, broad, expansive tail. Cf. juv. with juv. Striped Cuckoo. SOUNDS: Song (spring–summer) far-carrying, typically 3 deliberate, haunting whistles with tinamou-like quality, the 1st 2 notes suggesting Striped Cuckoo, the last variably quavering, *whee whee wheerr*; less often longer series, accelerating overall and slightly rising, *whee-whee whee'whee-bee*, and variations. STATUS: Scarce and local. (Mexico to S America.)

SCARLET MACAW,
TROPICAL CUCKOOS

SCARLET MACAW

COMMON
SQUIRREL CUCKOO

juv.

STRIPED CUCKOO

juv.

PHEASANT CUCKOO

YELLOW-BILLED CUCKOO *Coccyzus americanus* 28–30.5cm. Transient migrant in varied wooded and forested habitats from rainforest to cayes and coastal scrub. Mainly at mid–upper levels in trees, often in canopy, where sluggish and easily overlooked. Hops and peers about on foliage for caterpillars, other invertebrates. Flight fast and direct, often slipping into cover and vanishing. Note clean white underparts, big white tail spots, bright rusty flash in wings, bright yellow on bill; cf. Black-billed and Mangrove Cuckoos. Juv. has less contrasting tail pattern than adult. Typically silent in Belize. **STATUS:** Fairly common (especially on cayes) to uncommon, late Aug–early Dec, Apr–early Jun. (Breeds N America and Mexico, winters S America.)

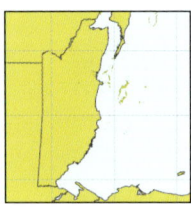

BLACK-BILLED CUCKOO *Coccyzus erythropthalmus* 27.5–30cm. Transient migrant in varied wooded and forested habitats from rainforest to cayes and coastal scrub. Habits much like Yellow-billed Cuckoo, thus overlooked easily and encountered infrequently. Note dark bill, rather plain head without dark masked look, dingy whitish underparts, weak pattern on underside of tail. Cf. Yellow-billed and Mangrove Cuckoos. Typically silent in Belize. **STATUS:** Scarce to uncommon Sep–Nov, mid-Apr to May, most records from cayes. (Breeds N America, winters S America.)

MANGROVE CUCKOO *Coccyzus minor* 30.5–33cm. Handsome cuckoo of mangroves, semi-deciduous forest and edge. Habits much like Yellow-billed Cuckoo, thus overlooked easily and encountered infrequently; like other cuckoos, often sits out and suns itself in early morning, at times on exposed branches. Note buff underparts, dark mask, big white tail spots, bright yellow on bill; cf. Yellow-billed and Black-billed Cuckoos. Juv. has less contrasting tail pattern. **SOUNDS:** Mostly silent unless breeding. Song a typically accelerating series of harsh croaks, 6–9 secs, ending with an abrupt switch to a quieter short series of lower coos, *AHRR, AHRR, AHRR-AAHR-AAHR…owh-owh…*; sometimes preceded by quiet dry rattle of 1–2 secs. Series of (usually 6–18) low nasal croaks in slower, steady series, *wohr wohr…*, 3–4/sec (vs. 4–7 notes/sec in song). Occasional short dry rattles at other seasons. **STATUS:** Scarce in coastal belt and cayes, mainly late Sep–Apr (may breed locally); rare and sporadic inland, mainly Oct–Apr. (Mexico to n. S America.)

GROOVE-BILLED ANI *Crotophaga sulcirostris* 30.5–34.5cm. The common ani in Belize, found in varied open and semi-open habitats, often around livestock. Usually in small groups, rarely to 30 birds. Perches on wires, hops readily on ground, clambers in foliage, long tail often loosely flopped about. Flight distinctive: rapid flaps interspersed with flat-winged glides, one bird following another across a field or road; often crash lands into foliage. Bill shape and habits distinctive, except for rare Smooth-billed Ani (which see). **SOUNDS:** Common call a squeaky *pí-chwiep* or *pí-weérp*, emphasis on 1st note, often in short series. Other varied piping, squealing, and growling noises. 'Song' heard infrequently, a rapid mellow clucking *whiuh-whiuh…*, typically several secs duration; might suggest a rapid Ferruginous Pygmy Owl. **STATUS:** Common to fairly common, including Ambergris Caye; wanders occasionally to inshore cayes, mainly in winter. (Mexico and s. Texas to S America.)

SMOOTH-BILLED ANI *Crotophaga ani* 33–37cm. Rare visitor to northern cayes. Habits much like Groove-billed Ani. Singles or small groups sometimes mix with Groove-billed Ani. Slightly larger than Grove-billed (noticeable in direct comparison), often best detected and identified by voice. Also note more raised culmen of Smooth-billed (especially male), smooth bill sides (can be hard to see, and beware juv. Groove-billed has smooth bill, lacking grooves for a few months after fledging). **SOUNDS:** Slurred squealing whistles, distinct from Groove-billed Ani, typically *reeéah* or *wheeéreh*, can suggest a hawk. **STATUS:** Scarce and sporadic on n. cayes, mainly Oct–Mar but records year-round; very rare visitor to n. mainland. (Caribbean region to S America.)

YELLOW-BILLED CUCKOO

juv.

BLACK-BILLED CUCKOO

MANGROVE CUCKOO

variation

GROOVE-BILLED ANI

male

SMOOTH-BILLED ANI

TROGONS (TROGONIDAE; 4 SPECIES) Pantropical family of fairly large, colorful, mostly forest-based birds. Ages/sexes differ; attain adult plumage in 1 year; juv. plumage soft and weak, soon molted, but pale-spotted upperwing coverts may be retained for much of 1st year, as is juv. tail, which has narrower feathers than adult tail, and often slightly different patterning. Nest in tree cavities, termite and wasp nests.

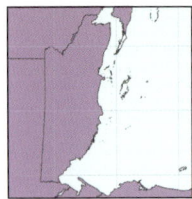

GARTERED [VIOLACEOUS] TROGON *Trogon caligatus* 24–25.5cm. Small, rather compact, yellow-bellied trogon of humid forest and edge; often nests in wasp nests. Mainly at mid–upper levels; often sings from subcanopy, at times on exposed perches, even roadside wires. Often occurs alongside appreciably larger, longer-tailed Black-headed Trogon; readily separated by structure, plumage, voice. **SOUNDS:** Song a fairly rapid, steady series (usually 4–9 secs) of overslurred nasal hoots, *kyow-kyow…* or *kuh-kuh…*, 3–4 notes/sec; easily mistaken at a distance for Ferruginous Pygmy Owl. Slightly overslurred dry growl, and series of nasal clucks, often with slightly laughing cadence. **STATUS:** Fairly common to common. (Mexico to nw. S America.)

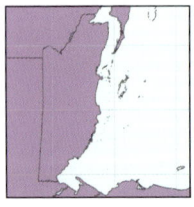

BLACK-HEADED TROGON *Trogon melanocephalus* 26.5–28cm. Medium-size, rather long-tailed trogon of humid forest edge, semi-open areas with scattered trees, hedgerows, mangroves; nests in occupied termitaries. Mainly at mid–upper levels, sometimes on roadside wires; birds gather at fruiting trees, where can be confiding. Singing males form 'leks' of up to10 birds. Often occurs alongside appreciably smaller, more compact Gartered Trogon, which has bold barring under tail (1st-year Black-headed has some dark bars at edges of tail); male Gartered also has yellow eyering, pale-gray wing panel; female has white eye-crescents, barred wing coverts. **SOUNDS:** Song a series of hollow nasal clucks that accelerate into a rattle, ends fairly abruptly, *kuh kuh-kuh… keh-keh…*, 2–4 secs. Calls include hollow, thrush-like clucks. **STATUS:** Fairly common to common on mainland, uncommon on Ambergris Caye, occasional in winter on Caye Caulker. (Mexico to nw. Costa Rica.)

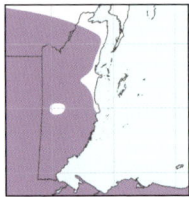

SLATY-TAILED TROGON *Trogon massena* 33–35.5cm. Large, stout-billed, red-bellied trogon of rainforest and edge, adjacent shady clearings with tall trees; often nests in termitaries. Mainly at mid–upper levels; sings mainly from subcanopy, often confiding. Distinctive, with stout orange bill, lack of white breast band, dark slaty undertail (narrowly barred white at edges on imm.). Only other red-bellied trogon in Belize is smaller Northern Collared Trogon. **SOUNDS:** Song a steady, often prolonged series of hard, overslurred clucks, *koh-koh…*, or *ka-ka…*, 3–4 notes/sec up to 30 secs or longer. Quiet clucking chatters, at times with laughing cadence. **STATUS:** Fairly common. (Mexico to nw. Ecuador.)

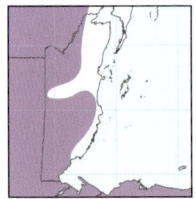

*NORTHERN COLLARED TROGON *Trogon [collaris] puella* 26.5–28.5cm. Relatively small red-bellied trogon of humid forest. Feeds and perches low to high, calls mainly from mid-levels in shady understory; can be fairly confiding. Widespread overlap with much larger, distinctive Slaty-tailed Trogon, the only other red-bellied trogon in Belize. **SOUNDS:** Song comprises 2–3 plaintive, downslurred whistles with measured cadence, *kyow kyow*, easily imitated, often rather quiet; less often *kyow kyow-kow* and rarely single notes or short series with slightly descending, laughing cadence. Call a slightly descending nasal growl, *ahrrrrr*, often repeated steadily as tail is raised and lowered. **STATUS:** Uncommon to fairly common, especially in hilly areas. (Mexico to Panama.)

GARTERED TROGON

imm.
male

female

males

BLACK-HEADED TROGON

female

males

SLATY-TAILED
TROGON

imm.
male

female

males

NORTHERN COLLARED
TROGON

females

males

148

KINGFISHERS (ALCEDINIDAE; 5 SPECIES) Worldwide family of small to fairly large birds with big heads, long pointed beaks. Ages differ slightly, like adult in 1st year; sexes differ. Nest in burrows in banks.

BELTED KINGFISHER *Megaceryle alcyon* 31–33cm. Widespread nonbr. migrant to varied habitats with water, from cayes and mangroves to lakes, rivers, roadside ditches; not usually in forested areas. Hunts from perches and from hovering fairly high over open water; often perches conspicuously on wires. Distinctive; cf. appreciably larger Tropical Ringed Kingfisher, which has rusty underparts. Slightly larger than Amazon Kingfisher but with smaller bill, distinct plumage, white wing patches. SOUNDS: Rapid-paced, dry, 'machine-gun' rattle, 1–5 secs, often in flight. STATUS: Fairly common Aug–Apr, rarely from Jul and into May. (Breeds N America, winters to nw. S America.)

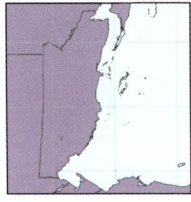

*TROPICAL RINGED KINGFISHER *Megaceryle torquata* 38–41cm. Mainly freshwater habitats, especially lakes, slow-moving rivers, less often mangroves, brackish water. Hunts from perches and from hovering fairly high over open water; often perches conspicuously on wires and regularly seen flying high overhead, sometimes well away from water. Flies with fairly slow, deep wingbeats. Distinctive, with solidly rusty underbody, large size, massive bill. Juv. resembles female but breast band darker, mottled cinnamon. SOUNDS: Deep *chrek!* in flight. Powerful chattering rattles, deeper, slower-paced, and often more prolonged than Belted Kingfisher. STATUS: Fairly common to common on mainland, occasional on Ambergris Caye. (Mexico and s. Texas to tropical S America)

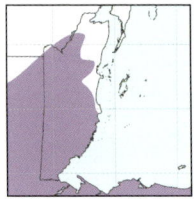

AMAZON KINGFISHER *Chloroceryle amazona* 28–29cm. Varied freshwater habitats, mainly in forested areas, from coastal lagoons and large rivers to small streams and ponds. Hunts mainly from perches, at times on wires, but tends to be less conspicuous than larger kingfishers; hovers occasionally, and flight usually low over water. Distinctive, with massive bill, dark oily-green upperparts; lacks bold white wing spotting of much smaller Green Kingfisher. Juv. resembles female but male upper breast washed buffy. SOUNDS: Gruff to low rasping *chruk* or *zzrk*, mainly in flight, at times run into rattling and screechy chatters. STATUS: Uncommon to fairly common; absent from north and cayes. (Mexico to S America.)

GREEN KINGFISHER *Chloroceryle americana* 19–21cm. Small darting sprite of fresh and brackish habitats, from small pools and streams to mangroves, larger rivers, lake edges. Hunts from perch, usually fairly concealed and low over water, rarely high on open wires; does not hover. Flight typically low and fast, flashing white outer tail feathers. Distinctive, with small size, contrasting white neck sides, distinct white wing spotting. Juv. resembles female but male upper breast washed buffy. SOUNDS: Dry rasping clicks, often run into short rattles; gruff buzzy *zzher* mainly in flight; short buzzy and squeaky chatters. STATUS: Fairly common to common; occasional reports from Ambergris and Caye Caulker. (Mexico and sw. US to S America.)

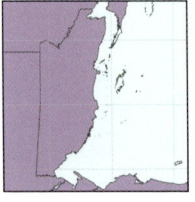

AMERICAN PYGMY KINGFISHER *Chloroceryle aenea* 13–14cm. Tiny darting sprite of fresh and brackish habitats, from forest pools and streams to mangroves; usually in wooded and forested areas, easily overlooked. Hunts from shady perch low over water; at times catches insects. Distinctive, with very small size, buffy neck sides, rusty underparts. SOUNDS: Dry ticking notes and short rattles, slightly higher and often softer than Green Kingfisher; high, slightly metallic burry *zzrieh*; downslurred, slightly squeaky to shrill short chatter. STATUS: Uncommon to fairly common on mainland; occasional reports from Ambergris and Caye Caulker. (Mexico to S America.)

BELTED KINGFISHER

male

female

TROPICAL RINGED KINGFISHER

male

female

AMAZON KINGFISHER

male

female

GREEN KINGFISHER

male

female

male

female

AMERICAN PYGMY KINGFISHER

MOTMOTS (MOMOTIDAE; 3 SPECIES) Small Neotropical family of large-headed, long-tailed forest birds. Ages differ slightly; juvs. duller overall but soon resemble adults; sexes similar. Newly molted tails are fully feathered, but larger species have intrinsically weakened sections soon removed by preening to produce 'racket tips.' Nest in burrows in banks.

LESSON'S [BLUE-CROWNED] MOTMOT *Momotus [momota] lessonii* 38–43cm. The common and widespread large motmot of Belize (but heard much more often than seen). Found in humid forest and edge, plantations; semi-open areas with larger trees, banks for nesting. Distinctive if seen well, but cf. Keel-billed Motmot of s. rainforest. Often sits quietly, easily overlooked if not vocal; switches tail side-to-side like a jerky pendulum. Perches low to high, often in subcanopy; regular at army ant swarms. **SOUNDS:** Song typically a low double hoot, *whoop woop*, less often short series or single notes; often given pre-dawn and easily mistaken for an owl. Calls include harsh low clucks and chatters, and a soft, bouncing-ball hooting series suggesting Spectacled Owl. **STATUS:** Fairly common to common. (Mexico to Panama.)

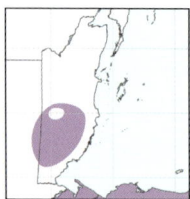

KEEL-BILLED MOTMOT *Electron carinatum* 31–33cm. Medium-size motmot of rainforest, especially hilly country near streams. Feeds low to high, but mainly at mid–upper levels inside shady forest, less often at edges. Calls mainly from subcanopy. Combination of turquoise-blue brow, black mask, and rusty forehead distinctive; also note relatively broad bill from below, cf. Lesson's Motmot. **SOUNDS:** Song a far-carrying, ringing nasal *kwaah* or *ownhh* every 3–6 secs. Varied rhythmic clucking series when excited. **STATUS:** Uncommon and local in Maya Mts. (Mexico to Costa Rica.)

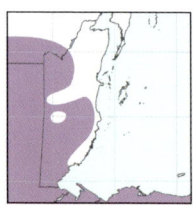

TODY MOTMOT *Hylomanes momotula* 16.5–18cm. Small, unobtrusive motmot of humid forest. Favors shady understory, from rather open situations to tangled gullies. Perches quietly at low to mid-levels, and flushes with low whirr of wings. Slowly flicks tail up and down, not side-to-side like larger motmots. Distinctive, but in shady dark understory cf. White-whiskered Puffbird. **SOUNDS:** Mainly in early morning, often before first light, song a nasal, slightly rising or overslurred hoot, *wah* or *woah*, usually in prolonged series, 10 notes/10–12 secs; can suggest Gartered Trogon but more nasal, usually slower-paced. Excited birds give faster series of burrier notes, 10 notes/2.5–6 secs, at times in pulsating duets, *wah'awah'awah….* **STATUS:** Uncommon to fairly common locally. (Mexico to nw. Colombia.)

PUFFBIRDS (BUCCONIDAE; 2 SPECIES) Neotropical birds of forest and forest edge. Ages/sexes similar or slightly different; attain adult appearance in 1st year. Nest in burrows.

WHITE-NECKED PUFFBIRD *Notharchus hyperrhynchus* 24–25.5cm. Distinctive, big-billed, boldly pied inhabitant of humid forest canopy. Seen mainly from edges or overlooks of canopy, as singles or pairs perched on high and often rather exposed branches of emergent, trees. Sits still and often quietly for long periods. **SOUNDS:** Song a high, slightly bubbling slow trill *wui-wui-wui…*, 3–8 secs; may suggest flight song of Great Black Hawk; at times ends with a few inflected nasal whistles, *k'wik k-wik…*or *wiki wiki….* **STATUS:** Uncommon to fairly common but low density. (Mexico to S America.)

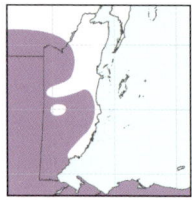

WHITE-WHISKERED PUFFBIRD *Malacoptila panamensis* 19–20cm. Inconspicuous, rather plump brown bird of rainforest understory and edge, adjacent shady clearings with trees. Singles or pairs perch quietly at low to mid-levels, where easily overlooked; can be quite confiding. Distinctive, but cf. Tody Motmot, which shares similar habits. Male rusty overall, female colder brown; extent of streaking below and pale spotting above variable. **SOUNDS:** Song a very high, penetrating, slightly descending reedy whistle, *tssiiiiiir*, about 1 sec, every 5–10 secs. Calls include very high, short, downslurred whistles and burry clucks. **STATUS:** Uncommon to fairly common. (Mexico to w. Ecuador.)

Lesson's Motmot

variation

Keel-billed
Motmot

Tody Motmot

White-necked
Puffbird

White-whiskered
Puffbird

male

female

JACAMARS (GALBULIDAE; 1 SPECIES) Neotropical family of forest and forest edge birds. Ages similar, sexes differ slightly; attain adult appearance in 1st year. Nest in burrows in banks.

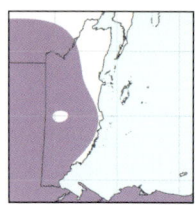

RUFOUS-TAILED JACAMAR *Galbula ruficauda* 22–23.5cm. Distinctive, slender, long-billed bird of rainforest edge, adjacent clearings and openings, especially with looping vines. Found as singles or pairs, typically perched at mid-levels on vines or slender branches, bill raised above horizontal. Sallies for flying insects and often returns to the same perch, where makes about-face leaps to switch the angle of its prey-seeking vigils. Male has white throat, female buff. **SOUNDS:** Loud calls often draw attention. High, shrieking, sharply overslurred *wheéuk*, often repeated steadily; may suggest Northern Royal Flycatcher. Faster, ringing series of whistles can end with a quick chortling trill, *whee-whee-whee…*. Song (?) a slightly rising then falling series of drawn-out whistles that accelerate into a short trill. **STATUS:** Uncommon to fairly common. (Mexico to S America.)

TOUCANS (RHAMPHASTIDAE; 3 SPECIES) Neotropical family of spectacular, big-billed, forest and forest edge birds most diverse in South America. Ages differ slightly (juvs. duller overall, with duller-patterned bills), attaining adult appearance in 1st year; sexes similar, but males average bigger and longer bills. Nest in tree cavities.

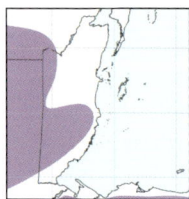

EMERALD TOUCANET *Aulacorhynchus prasinus* 32–37cm. Only green toucan in Belize, found in foothill rainforest, pine-evergreen forest, adjacent clearings; rarely ranging to lowland forest. Usually in pairs or small groups, often at fruiting trees, but easily overlooked given its green coloration. Flight fairly fast and direct with whirring wingbeats. **SOUNDS:** Low, grunting, often slightly burry *wuhk…* or *rruk…*, repeated steadily, 10 notes/3.5–6 secs; series can start with 1 or more well-spaced, low throaty growls, *aahrrr*. At a distance might suggest Keel-billed Toucan, which has higher, slightly longer, and creakier notes. **STATUS:** Fairly common in Maya Mts., rarely wandering to adjacent lowlands in winter; scarce in northwest, rarely wandering e. of mapped range. (Mexico to nw. Nicaragua.)

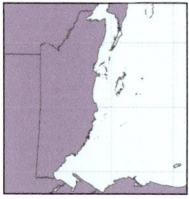

COLLARED ARACARI *Pteroglossus torquatus* 38–43cm. Distinctive smaller toucan of varied wooded and forested habitats, adjacent clearings with larger trees. No similar species in Belize. Typically in small groups, moving through canopy or flying across clearings one at a time, with rather direct flight and rapid wingbeats. Juv. has smaller and duller bill without strong black-edged serrations, yellowish facial skin soon becomes grayish, then red. **SOUNDS:** Sharp, squeaky, slightly metallic *pí-chi* or *skweí-zi*, sometimes repeated steadily; may suggest Groove-billed Ani. **STATUS:** Fairly common to common. (Mexico to nw. S America.)

KEEL-BILLED TOUCAN *Ramphastos sulfuratus* 51–59cm. The national bird of Belize. Unmistakable, with huge, rainbow-colored bill. More often heard than seen, but can be conspicuous, especially in early to mid-morning, perched on emergent bare snags in canopy or along forest edge; typically in pairs or small groups. Flight often high and deeply undulating, bursts of wingbeats interspersed with long swooping glides. Male has longer bill than female, noticeable within pairs. **SOUNDS:** Loud, slightly creaky croak, typically repeated steadily *rrek-rrek…*, 10 notes/4.5–7 secs; at a distance sounds like frogs (and cf. Emerald Toucanet), up close has an ear-splitting, shrieking quality. **STATUS:** Fairly common to common. (Mexico to nw. S America.)

RUFOUS-TAILED JACAMAR

female

male

EMERALD TOUCANET

COLLARED ARACARI

juv.

KEEL-BILLED TOUCAN

WOODPECKERS (PICIDAE; 11 SPECIES) Popular widespread family, absent Australasia. Ages/sexes differ slightly to distinctly; usually attain adult appearance in 1–2 months after fledging. Male usually has more red on head than female. Calls often useful for ID. 'Song' is mechanical drumming of bill on wood.

GENUS *CENTURUS* (3 species). Widespread New World group, often in edge and fairly open habitats with larger trees and hedgerows. Most species have black-and-white barring on back. Churring and rattled calls can be similar between species. Often merged into genus *Melanerpes*.

*GOLDEN-FRONTED WOODPECKER *Centurus (Melanerpes) aurifrons* 21.5–23cm. Familiar, conspicuous woodpecker of semi-open areas with taller trees, hedgerows, forest edge, woodland, gardens. Appreciably larger and bigger-billed than Yucatan Woodpecker with narrower white barring above creating darker overall appearance; red nasal tufts often stand out more strongly than yellow of Yucatan; voices distinct. Female has grayish crown, male has red crown merged into nape patch. Population on Turneffe Is. has more orangey-red crown and belly, broader white bars on wings. SOUNDS: Rapid-fire, rhythmic laughing *heh'eh'eh* or *heh-heh'eh'eh*. Varied, often rather gruff clucks, commonly doubled into a sneezy, *chuh-uh* or *cheh-eh*, 2nd note sometimes slurred, *cheh'ehrr*. Drum a hollow rattled *durrrr…*, about 1 sec. STATUS: Common and widespread. (Texas to nw. Nicaragua.)

YUCATAN (RED-VENTED) WOODPECKER *Centurus (Melanerpes) pygmaeus* 17–18cm. Small, perky woodpecker of semi-deciduous forest, beach scrub, clearings in humid forest, gardens. Regularly found alongside Golden-fronted Woodpecker, at times in the same tree. Often rather active, in bushes and on smaller twigs, as well as on larger trunks. Note small size, short bill; yellow around bill base can be difficult to see; broader white barring creates paler, more silvery back than Golden-fronted. Female has grayish crown, male has red crown merged into nape patch. SOUNDS: Nasal chuttering calls and laughing chatters; softer, higher, and faster-paced than Golden-fronted. Drum a rattled *darrrr…*, about 1 sec, higher than Golden-fronted. STATUS: Uncommon to locally fairly common in Corozal, uncommon and local to south, where appears to be spreading, at least in west; also on Ambergris and Caye Caulker. (Mexico to n. Belize, Honduras Bay Islands.)

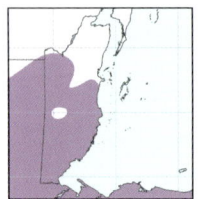

BLACK-CHEEKED WOODPECKER *Centurus (Melanerpes) pucherani* 18–19cm. Distinctive small woodpecker of rainforest, especially edges and clearings; feeds low to high on trunks, branches, at fruiting and flowering trees. Often in same areas as slightly larger Golden-fronted Woodpecker. Broad black mask distinctive, also note unbarred shoulders, cf. other *Centurus*. Female has red limited to nape patch, bordered black. SOUNDS: Rapid-fire, slightly nasal *huh'duh'duh*, often in short series; averages slower, burrier than similar call of Golden-fronted Woodpecker. Drum a rather hollow *durrrr…*, 0.5–1 sec, similar to Golden-fronted Woodpecker. STATUS: Fairly common to uncommon. (Mexico to w. Ecuador.)

LADDER-BACKED WOODPECKER *Dryobates (Picoides) scalaris* 14–15cm. Small woodpecker of lowland pine savanna, open and semi-open country with hedgerows, ranging to mangroves, clearings in humid forest. No visually similar species in Belize, but some calls might suggest Smoky-brown Woodpecker. Male has extensively red crown, spotted white; juv. (both sexes) has variable red crown patch. SOUNDS: Sharp *piik* or *chiik*, and rough, slightly descending, chattering rattle that fades out at end, 1.5–2 secs. Drum a relatively high, fast-paced *dirrrr…*, 1–1.5 secs, faster-paced than *Centurus* drums. STATUS: Uncommon to fairly common locally, especially in coastal belt; may be spreading locally with deforestation. (Mexico and sw. US to nw. Nicaragua.)

GOLDEN-FRONTED WOODPECKER

female

male

YUCATAN WOODPECKER

male

female

BLACK-CHEEKED WOODPECKER

male

female

LADDER-BACKED WOODPECKER

male

female

SMOKY-BROWN WOODPECKER *Dryobates (Veniliornis) fumigatus* 15–16cm. Small brown woodpecker of humid forest and edge, adjacent second growth. Forages low to high, often in vine tangles or other thick vegetation. Distinctive, but in dark shady understory cf. Golden-olive Woodpecker, which has paler face, barred underparts. Male has mottled red crown; juv. resembles male but female has red only on forecrown. **SOUNDS:** Sharp *chik!* and rough shrieky rattle, 2–3 secs; squeaky rhythmic *chwíka chwíka…* in interactions. Drum rapid and rather low, *durrrr…*, <1 sec. **STATUS:** Uncommon to fairly common. (Mexico to S America.)

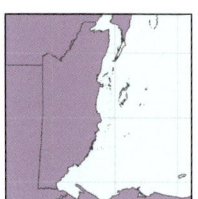

GOLDEN-OLIVE WOODPECKER *Colaptes rubiginosus* 21–22.5cm. The only green woodpecker in Belize, found widely in varied forested habitats, adjacent clearings with tall trees. Mainly at mid–upper levels in taller trees, where can remain still and quiet for long periods; easily overlooked unless vocal. Male has red mustache; juv. resembles adult but male's mustache mottled gray. **SOUNDS:** Rapid-paced, shrill churring rattle, 2–3.5 secs. Sharp, slightly explosive *keéah* or *kyaah*, recalling Northern Flicker *Colaptes auratus* of North America. Drum relatively low, moderate-paced, *urrrr…*, 1–2 secs. **STATUS:** Uncommon to fairly common; most numerous in Mountain Pine Ridge. (Mexico to S America.)

CHESTNUT-COLORED WOODPECKER *Celeus castaneus* 21.5–24cm. Stunning crested woodpecker of humid forest, adjacent clearings with larger trees. Feeds mainly at mid–upper levels, including at fruiting and flowering trees. No similar species in Belize; note wobbly crest, overall chestnut coloration with paler head. Male has broad red mustache; juv. malar mottled dusky. **SOUNDS:** Slightly explosive, overslurred hollow *whéow*, at times followed by a short cluck, *kéyow hik*, or by short laughing series of nasal notes; sharp, nasal, slightly squeaky *wíchk!* Drum relatively high and hollow, often rather soft, *ohrrrr…*, 1–1.5 secs. **STATUS:** Uncommon to fairly common. (Mexico to w. Panama.)

YELLOW-BELLIED SAPSUCKER *Sphyrapicus varius* 19–20.5cm. Winter migrant to varied open forested habitats, semi-open areas with taller trees, hedgerows, gardens; not in heavy rainforest. Feeds low to high on trunks and larger branches; often rather sluggish, easily overlooked. Presence often revealed by 'sapsicles'—neat rows of holes drilled on trunks to access sap. Male has red throat, female white; some females have black crown. Retains much juv. plumage (mottled dirty pale brownish) into winter, resembles adult by Mar–Apr. **SOUNDS:** Mostly quiet. Mewing downslurred *meeah* mainly in interactions. **STATUS:** Uncommon mid-Oct to Mar (sometimes from late Sep and into Apr), including cayes. (Breeds N America, winters s. US to Panama.)

SMOKY-BROWN WOODPECKER

male

female

GOLDEN-OLIVE WOODPECKER

male

female

male

imm.

YELLOW-BELLIED SAPSUCKER

male

CHESTNUT-COLORED WOODPECKER

female

female

ACORN WOODPECKER *Melanerpes formicivorus* 21–23.5cm. Distinctive, social, and often noisy 'clown-faced' woodpecker of pine savanna and pine forest intermixed with oaks. Feeds from ground to canopy; specializes in acorns, which are stored in holes drilled in 'granary trees.' Often conspicuous, usually in pairs or small groups; sometimes sallies and soars for flying insects. Male lacks black forehead band; juv. (both sexes) resembles male but eyes dusky, soon like adult. **SOUNDS:** Varied nasal laughing and crowing calls, including rhythmic *yáka yáka yáka…*; rolled churring *prrreh* and *krreh'eh*. Drum relatively slow-paced, about 1 sec, often slightly slower at start and end. **STATUS:** Fairly common to common locally; occasional wanderer outside of mapped range. (N America to n. Colombia.)

LINEATED WOODPECKER *Dryocopus lineatus* 31.5–34cm. Large 'woody woodpecker' of varied forested and semi-open habitats, mangroves, hedgerows, villages. Feeds low to high, and sometimes associates with Pale-billed Woodpecker, even nesting in the same tree. Pale-billed slightly larger with blockier red head, white lines on back more closely spaced into broken V, not parallel lines; beware juv. Pale-billed, which has black-and-white face recalling Lineated. Male has red mustache. **SOUNDS:** Fairly rapid, overall steady, laughing series of yelping clucks, *yeh-yeh-yeh…*, often gets louder and then fades abruptly at end, 3–6 secs; recalls Northern Flicker of N America. Sharp chik notes, often repeated steadily or run into a low growl, *puik! errrr*, recalling Common Squirrel Cuckoo. Drum powerful and resonant, *Durrrr…*, fairly slow-paced, 1–2 secs. **STATUS:** Fairly common to common. (Mexico to S America.)

PALE-BILLED WOODPECKER *Campephilus guatemalensis* 35.5–38cm. Largest woodpecker in Belize, found in forested habitats with large trees. Habits much like Lineated Woodpecker, but Pale-billed is more of a forest-based bird, not likely to be found in semi-open areas, towns, and villages. Note close-spaced white back stripes forming a V. Adult male has red head, female has black forehead and throat. Juv. (late winter through summer) very different, with black face and white stripe recalling Lineated Woodpecker, attains adult appearance in 1–2 months. **SOUNDS:** Sharp nasal clucking, often repeated persistently with hesitant cadence; recalls a squirrel scolding. Drum a distinctive, far-carrying, rapid double-rap, *D-DUK*; sometimes fades into a few slower-paced taps. **STATUS:** Fairly common. (Mexico to w. Panama.)

ACORN WOODPECKER

male

female

LINEATED WOODPECKER

male

female

PALE-BILLED WOODPECKER

male

female

juv.

JAYS (CORVIDAE; 3 SPECIES) Worldwide family (including crows) of social, intelligent, and often noisy birds found mainly in wooded and forested habitats. Ages differ slightly (to distinctly in Yucatan Jay); attain adult plumage in 1st year, but can retain patches of imm. bill color into 2nd year; sexes similar.

YUCATAN JAY *Cyanocorax yucatanicus* 32–34.5cm. Striking blue-and-black jay of semi-deciduous forest and edge, gallery forest in pine savanna, clearings with scattered trees and shrubs, plantations. No similar species in Belize. Forages low to high, usually in small groups, at times flocks of 20 or more. Associates readily with other jays, orioles; attends army ant swarms. Juv. has whitish head and body, white-tipped tail; attains black head and body plumage by fall; yellow bill and eyering become dark in 2nd year. **SOUNDS:** Varied rattling staccato chatters; drier than Green Jay and can suggest orioles but rougher, vary from short and abrupt to persistent, more measured, up to several secs; also quick, rolled, slightly metallic low *chuh-chuht*; ringing metallic *jihnk jihnk* similar to Green Jay; loud, ringing, nasal *ch'ik ch-chik*. **STATUS:** Common in Corozal, becoming more local and less numerous southward in coastal belt, where exceptional wanderer s. of mapped range; very local in w. Orange Walk and nw. Cayo. (Mexico to Guatemala and Belize.)

BROWN JAY *Psilorhinus morio* 38–43cm. Very large jay of open forest and edge habitats, open and semi-open areas with hedgerows and taller trees. Usually in loose groups of 5–15 birds; forages low to high, from forest canopy to out in open grassy fields. Like most jays, shifts easily from loud and obnoxious to quiet and relatively elusive. Flight rather slow and fairly direct; flying across open areas at a distance can suggest Montezuma Oropendola. 1st-year has yellow bill and eyering, becoming dark in 2nd-year. **SOUNDS:** Notably limited vocabulary. Monotonous and all-too-soon familiar, a loud screaming *KYEEAH!* or *KYAAH!* often repeated mercilessly; accompanied by quiet explosive pop, audible at close range; can suggest Red-shouldered Hawk *Buteo lineatus* of North America. Also a more mewing *reyaah*, which may be repeated steadily, mainly in breeding season. **STATUS:** Common to fairly common. (Mexico to nw. Panama.)

*GREEN JAY *Cyanocorax luxuosus* 26–29cm. Distinctive small jay of humid forest and edge, pine forest, adjacent clearings. Usually in small groups that can be alternately noisy and shy, mainly at low to mid-levels; at times join mixed flocks with other jays, orioles; yellow tail sides often flash in flight. No similar species in Belize; yellow tail sides often flash in flight. Juv. duller overall, with dark eyes. **SOUNDS:** Commonly a rough rasping *jehr*, typically in short chatters or as prolonged persistent scolding; and bright ringing clucks or yelps, typically doubled or in short series, such as *yink-yink, yink-yink-yink*. Also a dry scolding *cheh-cheh...*; buzzy nasal *jehr jihjihjih*; throaty frog-like croaks and dry rattles. **STATUS:** Uncommon to locally fairly common, especially away from coast; most numerous in Mountain Pine Ridge. (Mexico and s. Texas to Honduras.)

YUCATAN JAY

juv.
(Jul–Sep)

1st-year

BROWN JAY

1st-year

GREEN JAY

HUMMINGBIRDS (TROCHILIDAE; 22 SPECIES) 'Hummers' are a distinctive New World family best known for their small size, spectacular flight powers, and brilliant colors. Ages differ slightly to strongly; sexes similar or different; adult appearance attained in 1st year. Species ID can be challenging in the field, when views are often brief and many birds get away as unidentified. Voice is often useful for ID and detection.

Feeders offer the best chances to see many species well, and will also help reveal local seasonal movements as well as turn up wandering individuals. Subtropical species in particular are prone to wander and might show up well outside their mapped ranges.

Many colors on hummingbirds are iridescent, and thus lighting plays a big role in how these appear in life. The plate illustrations show how colors would ideally look if perfectly lit, but this is rarely the case. The main problems lie with throat patches (known as gorgets) and crown patches, which often appear dark or colorless, and with tail coloration, which can change appreciably with only a slight change in light angle.

HERMITS (3 species). Mainly forest-based hummers with long arched bills, striped faces, and graduated tails that on some species have long white central feathers. Ages/sexes mostly similar. Mainly feed low and often 'trap-line,' making predictable circuits to isolated single flowers or groups of flowers (like a hunter checking traps); sometimes defend flower patches in *Heliconia* thickets.

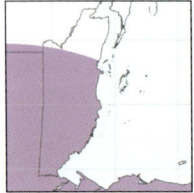

LONG-BILLED HERMIT *Phaethornis longirostris* 16–17cm. Large, spectacular if dull-plumaged hummer of rainforest understory, adjacent second growth, *Heliconia* thickets. Note long white tail streamers , long arched bill, striped face, pale grayish underparts. Typical views are brief, as a bird zips between widely spaced flowering patches to feed, often hovering briefly with white tail streamers near vertical and quivered. Singing males gather seasonally in leks, perching 1–4m up and often 10–15m apart in shady understory; tail wagged constantly while singing. Ages/sexes similar. **SOUNDS:** Call a high, lisping, emphatic *sweik!* often given in flight as birds flash by. Song a monotonously repeated, slightly buzzy *zzreih, zzreih…*, about 2 notes/sec. **STATUS:** Fairly common to common. (Mexico to Colombia.)

STRIPE-THROATED HERMIT *Phaethornis striigularis* 9–9.5cm. Tiny, distinctive hummer of rainforest and edge, *Heliconia* thickets. Note striped face, rusty rump. Feeds low to high, zipping quickly among and between flowers, often low along edges. Males gather to sing, usually a few birds 5–10m apart, perched low in dense understory, often within 1m of the forest floor; tail wagged constantly while singing. Ages/sexes similar. **SOUNDS:** Call a high sharp *siip!* suggesting Long-billed Hermit but weaker. Song a high, squeaky, prolonged jerky warble with slightly tinny quality, often frustratingly difficult to pinpoint; suggests song of Orange-billed Sparrow but usually slower-paced, more hesitant. **STATUS:** Fairly common to common. (Mexico to nw. S America.)

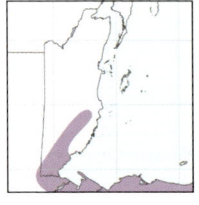

BAND-TAILED BARBTHROAT *Threnetes ruckeri* 11–12cm. Attractive hermit of s. rainforest and edge, *Heliconia* thickets. Mainly low and often difficult to see well, zipping quickly in shady understory; males sing alone or in small leks, from low perch in forest understory. Note distinctive face and breast pattern, white base to tail. Sexes similar; juv. duller overall. **SOUNDS:** High, sharp, down-slurred *ziik*, often doubled, shriller than other hermit calls. Song alternates descending squeaky trills and high downslurred notes; has overall slightly jerky cadence, up to 7 secs. **STATUS:** Scarce to uncommon in lower foothills and adjacent lowlands of south. (Guatemala to nw. S America.)

LONG-BILLED STARTHROAT *Heliomaster longirostris* 11.5–12.5cm. Very rare but striking, long-billed hummer of humid lowland forest edge, plantations, adjacent clearings and gardens with flowering bushes. Note face pattern, very long straight bill, white back patch. Crown iridescent turquoise on male, blue-green on most females. Imm. has little or no red in throat. **SOUNDS:** Sharp *chiup* while feeding and hovering. **STATUS:** Very rare, irregular visitor at any season to s. lowlands and lower foothills; most records during May–Jul. (Mexico to S America.)

Long-billed Hermit

variation

Stripe-throated Hermit

juv.

Band-tailed Barbthroat

Long-billed Starthroat

female

male

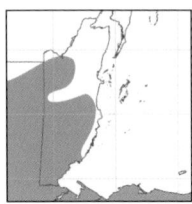

WHITE-NECKED JACOBIN *Florisuga mellivora* 11–12cm. Large flashy hummer of rainforest edge, plantations, adjacent habitats with flowering trees. Mainly at mid–upper levels, hovering in fairly horizontal plane with tail slightly cocked, at times flashed open. Male (some females similar) striking and distinctive, with blue hood, extensively white tail; female best identified by chunky shape, heavy black bill, scalloped bib, and white belly, cf. Scaly-breasted Hummingbird. Imm. male has buff on sides of throat. SOUNDS: Rather quiet. High, slightly wiry chips and twitters on occasion. STATUS: Uncommon to fairly common. (Mexico to S America.)

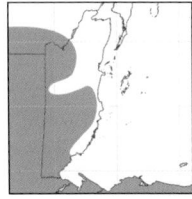

SCALY-BREASTED HUMMINGBIRD *Phaeochroa cuvieri* 11.5–12.5cm. Large hummer of rainforest edge, adjacent clearings and gardens with taller trees, plantations. Feeds low to high, often at mid–upper levels; sings mostly from high, rather open perches. Note drab plumage, medium-length black bill, white eye-spot, bold white tail corners, dingy buff belly; 'scaly breast' rarely striking. Cf. female White-necked Jacobin. Ages/sexes similar. SOUNDS: Song a rather loud, fairly slow-paced chanting of varied chips, squeaks, and thin whistles with slightly jerky cadence, often prolonged; might suggest a euphonia. Call a sharp chip, recalling Yellow Warbler. STATUS: Uncommon to fairly common. (Mexico to Colombia.)

PURPLE-CROWNED FAIRY *Heliothryx barroti* 11.5–13cm. Flashy hummer of rainforest and edge. Mainly at mid–upper levels where feeds actively, dashing from flower to flower (often pierces flower bases) or darting for insects, often flashing its tail open; rarely seen perched. Distinctive, with gleaming white underparts, short pointed bill, black mask, white-sided tail (much longer on female). Male has violet crown; juv. has sparse dark spotting on breast, cinnamon edgings to upperparts. SOUNDS: Mostly quiet; high, slightly metallic *ssit*, at times run into twitters. STATUS: Uncommon to fairly common. (Mexico to w. Ecuador.)

GREEN-BREASTED MANGO *Anthracothorax prevostii* 11–11.5cm. Large stocky hummer of open and semi-open areas with taller trees and hedgerows, gardens, forest edge. Often perches (and nests) conspicuously on tall bare branches and twigs. Feeds low to high, and flashes tail open when hovering. Distinctive, with rather thick, arched black bill, bright purple to coppery purple in tail; dark median stripe on female/imm. underparts; some adult females resemble adult male. Imm. has rusty mottling on sides of throat and breast. SOUNDS: Often quiet. High sharp chips and twitters on occasion. STATUS: Fairly common on cayes and in many coastal areas, uncommon and more local inland; seasonal movements complex, and in some areas (e.g., Toledo) less numerous or locally absent Nov–Jan. (Mexico to Panama.)

WHITE-NECKED JACOBIN

imm. male

'typical' female

adult males (some females similar)

SCALY-BREASTED HUMMINGBIRD

PURPLE-CROWNED FAIRY

female

males

GREEN-BREASTED MANGO

imm.

female

adult males (some females similar)

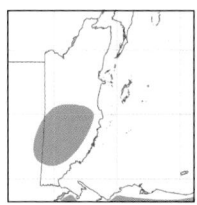

VIOLET SABREWING *Campylopterus hemileucurus* 14–15cm. Spectacular, very large hummer of humid foothill forest and edge, adjacent second growth. Feeds low to high; sings mainly from mid-level perches in shady subcanopy. Distinctive, with thick arched bill, white eyespot, big white tail corners often flashed in flight. Male can look simply dark overall, but violet tones stunning in the right light; often rather aggressive, chasing off almost all other hummers. **SOUNDS:** Hard sharp chips at times run into rattles. Song of varied sharp chips and short warbles often punctuated with fairly shrill, slightly explosive notes. **STATUS:** Uncommon to fairly common at higher elevations in Maya Mts.; uncommon and local in Mt. Pine Ridge; rare and sporadic (seasonal?) at lower elevations. (Mexico to w. Panama.)

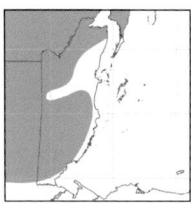

***WEDGE-TAILED SABREWING** Pampa (Campylopterus) curvipennis* 12–13cm. Distinctively large but plain, long-tailed hummer of humid forest and edge, adjacent clearings and second growth. Feeds low to high; sings mainly from mid-level perches, often in vine tangles and shady subcanopy. Distinctive if uncolorful, with stout bill, white eye-spot, pale grayish underparts, long graduated tail (longer on male, tipped whitish on female); both sexes have iridescent violet-blue crown. Imm. has buffy tips to upperparts, buff tinge to pale tail tips. **SOUNDS:** Sharp rich *chiup*, vaguely suggesting Kentucky Warbler, run into prolonged chatters or rattles when excited; high sharp *peek*. Song can be arresting: a loud, prolonged, gurgling warble interspersed with squeaky chipping; typically starts with hesitant, reedy chippering that may go on a min or longer before breaking into full song. **STATUS:** Uncommon to fairly common. (Mexico to Honduras.)

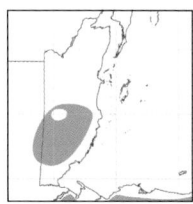

***STRIPE-TAILED HUMMINGBIRD** Eupherusa eximia* 9.5–10cm. Medium-size hummer of s. cloud forest and humid foothill forest. Feeds and perches mainly at low to mid-levels in shady understory. Feeding flight often quick and darting, hard to observe clearly. Distinctive in range, with rusty wing panel, white tail flashes (inner webs of outer 2 rectrices, often striking when hovering, concealed from above when tail closed). On female, also note plain, pale gray face and underparts, lack of white postocular spot, black bill. **SOUNDS:** High, slightly liquid rolled chips and pittering short trills when hovering; suggests Costa's Hummingbird *Calypte costae* of sw. US. Song a rapid, high, slightly liquid, chipping warble. **STATUS:** Fairly common at higher elevations in Maya Mts.; uncommon to scarce (seasonal?) at lower elevations. (Mexico to w. Panama.)

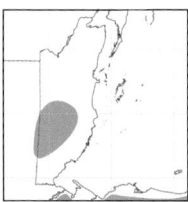

BROWN VIOLETEAR *Colibri delphinae* 10.5–11.5cm. Drab but subtly attractive hummer of humid s. foothill forest and edge, adjacent second growth. Mainly at mid–upper levels, often in fairly open canopy; singing males form loose groups. No similar species in Belize: note brownish plumage, broad whitish mustache, short bill, broad cinnamon edgings to rump feathers. Sexes similar. **SOUNDS:** Song from perch a series of (usually 3–8) strong, slightly metallic downslurred chips, *tchik tchik...*, repeated after short pause. Rattling chatters in interactions. **STATUS:** Uncommon to fairly common locally at higher elevations in Maya Mts. Very rare and sporadic (seasonal?) at lower elevations. (Belize to S America.)

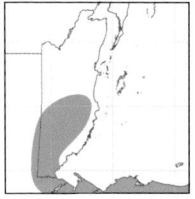

CROWNED WOODNYMPH *Thalurania colombica* 10–11cm. Medium-size hummer of s. rainforest, adjacent second growth and plantations. Feeds low to high, often in shady understory. Male distinctive, stunning when colors catch the light, with glittering emerald and purple plumage, forked blue-black tail. Female distinctive, with plain whitish underparts, small white corners to blue-black tail. Imm. male resembles female but with emerald patches on breast, tail lacks pale corners. **SOUNDS:** Dry, fairly hard rattling chips, at times run into rapid chatters. **STATUS:** Uncommon to scarce and local in south. (Belize to nw. S America.)

VIOLET SABREWING

female

males

imm.
female

**WEDGE-TAILED
SABREWING**

**STRIPE-TAILED
HUMMINGBIRD**

female

male

females

males

female

**BROWN
VIOLET-EAR**

CROWNED WOODNYMPH

males

female

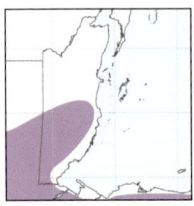

BLACK-CRESTED COQUETTE *Lophornis helenae* 6.5–7cm. Scarce tiny hummer of humid forest and edge, adjacent clearings with flowers, mainly in foothills. Feeds low to high, from canopy of flowering trees such as *Inga* to roadside flower banks. Flight often slow and deliberate, with tail cocked; males perch on bare twigs in canopy or subcanopy. Distinctive: with broad white rump band, as likely to be mistaken for an insect (some large moths deceptively similar) as for any other hummer in its range; also note coarse bronzy spotting on underparts, short reddish bill. Imm. male plumage variable, intermediate between female and adult male. SOUNDS: Mostly quiet. Soft chips and high twitters sometimes when feeding. Song from perch a clear, upslurred *tsuwee*, repeated. STATUS: Scarce in s. foothills and adjacent lowlands. (Mexico to Costa Rica.)

*CANIVET'S EMERALD *Cynanthus (Chlorostilbon) canivetii* Male 8.5–9cm, female 7.5–8cm. Very small hummer of humid forest edge, overgrown clearings, scrubby second growth, pine savanna, gardens. Feeds mainly at low to mid-levels, often wagging and fanning tail as it hovers. No similar species in Belize; note small size, tail-wagging behavior, female face pattern. Imm. male resembles female with longer tail, variable emerald green mottling on underparts. SOUNDS: Dry, rather staccato chatters; may suggest Ruby-crowned Kinglet *Corthylio calendula* of North America. STATUS: Fairly common to uncommon. (Mexico to Belize and Honduras Bay Islands.)

RUBY-THROATED HUMMINGBIRD *Archilochus colubris* 8–9cm. Small migrant hummer of varied open and semi-open habitats, from coastal scrub to rainforest edge, hedgerows, weedy fields. Feeds low to high, often in flowering roadside trees. Nothing really similar in Belize; adult male gorget can appear all-dark depending on light angle; on female note black-and-white tail corners, cf. White-bellied Emerald. SOUNDS: High twangy *tchi*, at times doubled, and varied twitters. STATUS: Uncommon to fairly common on mainland Oct–Apr (a few from mid-Sep and rarely into early May); most numerous Feb–early Apr; scarce transient on cayes. (Breeds N America, winters Mexico to w. Panama.)

WHITE-BELLIED EMERALD *Chlorestes (Amazilia) candida* 9–9.5cm. Small white-bodied hummer of humid forest and edge, plantations. Feeds low to high; sings mainly from bare twigs in subcanopy. Note small size, pinkish mandible, greenish tail with dark subterminal band but no black-and-white corners. Cf. Azure-crowned Hummingbird, female Ruby-throated Hummingbird. Ages/sexes similar. SOUNDS: High rolled and trilled chips and twitters; not as hard and rattled as Rufous-tailed Hummingbird. Song a short, typically jerky series of high, slightly shrill chips, repeated. STATUS: Fairly common to common. (Mexico to n. Nicaragua.)

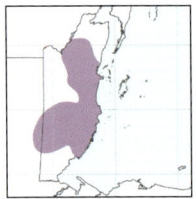

AZURE-CROWNED HUMMINGBIRD *Saucerottia (Amazilia) cyanocephala* 10–11.5cm. Medium-size hummer of pine-oak savanna and cloud forest, also locally in humid lowland forest and edge. Feeds low to high, often at mid–upper levels in trees. Note blue-green hood and dusky chest sides contrasting with white throat bib, white median underparts; dull bronzy-green rump and plain tail. Cf. White-bellied Emerald. Imm. has dull crown, pale grayish tail tips. SOUNDS: Sharp, slightly nasal chips often run into short rolls or trills; fairly hard buzzy *dzzzrt* in flight and repeated from perch. STATUS: Fairly common to uncommon but often rather local. (Mexico to Nicaragua.)

males

BLACK-CRESTED COQUETTE

female

CANIVET'S EMERALD

female

males

RUBY-THROATED HUMMINGBIRD

female

males

WHITE-BELLIED EMERALD

AZURE-CROWNED HUMMINGBIRD

CINNAMON HUMMINGBIRD *Amazilia rutila* 10–11.5cm. Fairly large, distinctive hummer of semi-deciduous forest and edge, gardens, mainly in drier and coastal areas. Feeds low to high, often fairly aggressive. Solidly cinnamon underparts distinctive; from behind, cf. Buff-bellied Hummingbird, which has green throat and breast. Sexes similar; imm. has mostly dark maxilla. SOUNDS: Fairly hard, slightly buzzy *tzk* and buzzy rattles. STATUS: Fairly common to common on n. coast (and on numerous cayes), in smaller numbers and more locally inland and s. along coast. (Mexico to nw. Costa Rica.)

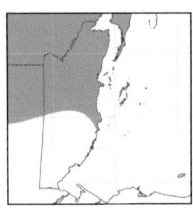

BUFF-BELLIED HUMMINGBIRD *Amazilia yucatanensis* 10–11cm. Medium-size hummer of semi-deciduous forest and edge, scrub, gardens. Feeds low to high, mainly at low to mid-levels. Rufous-tailed Hummingbird favors more humid habitats; note brighter belly of Buff-bellied, similar in tone to undertail coverts (rusty undertail coverts contrast distinctly with duller belly on Rufous-tailed), broader green edging to more forked tail than Rufous-tailed; lacks rufous lores of Rufous-tailed. From behind, cf. Cinnamon Hummingbird. Sexes similar; imm. has mostly dark maxilla, duller underparts. SOUNDS: Clipped, slightly smacking *tik*, at times run into rattles, not as hard as Rufous-tailed Hummingbird. In apparent display, makes wide-ranging, fast erratic flights while giving high sharp *siik!* calls. STATUS: Fairly common to common in north, less numerous at s. edges of range. (Mexico and s. Texas to Belize.)

RUFOUS-TAILED HUMMINGBIRD *Amazilia tzacatl* 10–11cm. Common, medium-size hummer of humid forest edge, adjacent second growth, gardens. Feeds low to high, mainly at low to mid-levels. Cf. Buff-bellied Hummingbird, which favors drier habitats. Male has red bill, iridescent throat and chest, dirty buff to buffy-gray belly; female has dark maxilla, duller throat and chest, pale grayish belly. SOUNDS: Fairly hard to sharp staccato chips from perch and when feeding; downslurred rattled trill when agitated, 1–2 secs. High rapid twitters in chases. Song a varied short series of high, thin, slightly squeaky notes. STATUS: Common to fairly common, less numerous in drier n. areas. (Mexico to w. Ecuador.)

BLUE-THROATED GOLDENTAIL *Chlorestes (Hylocharis) eliciae* 8–9cm. Very rare. Small hummer of humid lowland forest and edge, gardens. Feeds low to high, from flower banks to canopy of flowering *Inga* trees; perches mainly at mid–upper levels in fairly open but shady subcanopy. Despite the colorful name, often appears rather dull. Note small size (smaller than Rufous-tailed Hummingbird), bright red bill, greenish-gold tail (often looks grayish). Male throat often looks grayish with dark or bluish mottling unless in the right light. Cf. Rufous-tailed Hummingbird. SOUNDS: High squeaky chips and twitters. Song a lisping note followed by a short chatter, can be similar to some songs of White-bellied Emerald. STATUS: Unclear: known from 3 Apr–May records in south; perhaps a sporadic seasonal visitor to s. Belize. (Mexico to n. Colombia.)

Cinnamon Hummingbird

Buff-bellied Hummingbird

Rufous-tailed Hummingbird

female

males

males

female

Blue-throated Goldentail

SWIFTS (APODIDAE; 6 SPECIES) Worldwide family of supreme aerialists, seen perched only at roosts and nests, which tend to be in caves, behind waterfalls, and in other rarely accessible places. Ages/sexes similar in most species; adult appearance attained in about 1 year.

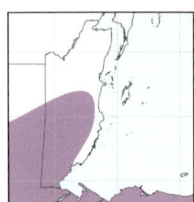

CHESTNUT-COLLARED SWIFT *Streptoprocne rutila* 12.5–14cm; WS 30.5–33cm. Poorly known small swift that appears dark overall unless seen in good light. Usually in small flocks (occasionally to 75+ birds), associating readily with other swifts. Flight recalls *Chaetura* swifts but stronger, less fluttery, with more frequent gliding and often higher overhead; note longer tail than *Chaetura* (typically notched), more scimitar-like wings that reflect more strongly silvery below, voice; chestnut collar diagnostic when visible. **SOUNDS:** Dry buzzy crackling notes and chatters, may suggest electricity crackling in power lines; occasional screechy notes thrown in. **STATUS:** Scarce to uncommon presumed resident, mainly in foothills, but may wander to s. coastal lowlands, especially in rainy weather. (Mexico to S America.)

LESSER SWALLOW-TAILED SWIFT *Panyptila cayennensis* 12.5–14cm; WS 29–31cm. Distinctive small swift of humid forest, adjacent areas such as riversides and over lagoons; note striking plumage pattern, long tail. Singles or small groups at times associate with flocks of Richmond's Swift; also flies higher, apart from Richmond's, and overlooked easily unless voice is learned. Flight very fast with flickering wingbeats, short glides, tail usually closed in a point. **SOUNDS:** Slightly explosive downslurred *spiez*, fading abruptly; reedy chips and buzzy twitters in interactions. **STATUS:** Uncommon to fairly common, mainly away from coastal belt; relatively scarce and local in Corozal. (Mexico to S America.)

WHITE-COLLARED SWIFT *Streptoprocne zonaris* 20–21.5cm; WS 48.5–53cm. Large spectacular swift ranging widely, mainly over forest and adjacent areas; nests colonially in caves and sinkholes, often near waterfalls. Feeds low to high, at times sweeping by at head height when wings make strong rushing sound; soars frequently and lazily, wings spread in paddle-like bulges. Usually in groups, locally up to a few 100 birds, from widely dispersed feeding bands to tightly synchronized screaming squadrons that wheel high overhead. Distinctive, the only large swift in Belize; note broad white collar, forked tail. 1st-year has solid white restricted to a broad hindcollar, with variable whitish scaling across upper neck. At a distance, single birds might be mistaken for Bat Falcon and vice versa. **SOUNDS:** Loud screaming and screeching chatters; noisy flocks can suggest parakeets; like other swifts, single birds typically silent. **STATUS:** Fairly common to common, mainly inland; rare wanderer n. of mapped range and unrecorded from cayes. (Mexico to S America.)

*WHITE-CHINNED SWIFT *Cypseloides cryptus* 14–15cm; WS 33–35.5cm. Poorly known; might be encountered anywhere in s. Belize, perhaps mainly in rainy weather. Slightly larger than Chestnut-collared Swift and chunkier, with bigger head, squared tail. With good views, note whitish forehead and chin (can be conspicuous in good light, especially when throat is full of food); female and juv. have variable whitish mottling on belly, which can appear as a whitish vent band. Wingbeats rapid, direct flight heavy-bodied, but soars readily. Associates with other swifts, especially Chestnut-collared, which has longer tail (notched on male), smaller head, often shows chestnut collar. **SOUNDS:** Sharp buzzy chips and chatters; more shrieky, less 'electric' or dry than Chestnut-collared. **STATUS:** Known definitely in Belize from 4 specimens collected in Aug 1931 at Manatee Lagoon, on central coast, and photos in Aug 2021 from w. Cayo; also provisional sight records in s. foothills. Perhaps simply a wanderer on feeding commutes from Honduras, but may be a scarce local resident, breeding in Maya Mts. (Honduras to S America.)

Chestnut-collared Swift

females

males

Lesser Swallow-tailed Swift

Richmond's Swift for comparison (see p. 174)

1st-year

White-collared Swift

adults

White-chinned Swift

*RICHMOND'S [VAUX'S] SWIFT *Chaetura [vauxi] richmondi* 10.5–11cm; WS 25–27.5cm. The common small swift in Belize. Usually seen in small groups, rarely up to 100 or so birds, less often singles; often feeds with other swifts, swallows. Flight usually relatively low over forest canopy, unlike larger swifts. Plumage much like larger and longer-winged Chimney Swift, which occurs in Belize only during spring and fall migration; larger size of Chimney readily apparent when seen alongside Richmond's but difficult to judge on lone birds. When in doubt, *Chaetura* sp. is a safe ID. **SOUNDS:** High thin chips and twitters, including an accelerating, overall descending *tsi-tsi-si-sirr*; higher and shriller than Chimney Swift. **STATUS:** Fairly common locally throughout, especially in hilly country; least frequent near coast and unrecorded from cayes. (Mexico to Cen America.)

CHIMNEY SWIFT *Chaetura pelagica* 12–12.5cm; WS 30–32.5cm. Transient migrant, much like smaller Richmond's Swift in plumage but rump averages duller. Best told by stronger, less fluttery flight (often rather direct, north in spring, south in fall), longer wings. Size difference readily apparent when seen alongside Richmond's but difficult to judge on lone birds; larger size and stronger flight of Chimney can recall Chestnut-collared Swift. Often in loose groups, also singles later in migration. **SOUNDS:** Usually silent. In interactions may give high smacking chips and short twitters, lower and fuller than Richmond's. **STATUS:** Fairly common Sep–early Nov, stragglers into late Nov; uncommon mid-Mar to Apr, a few into May; most frequent near coast and on cayes, but could occur anywhere. (Breeds e. N America, winters S America.)

SWALLOWS (HIRUNDINIDAE; 10+ SPECIES) Worldwide family of aerial insectivores. Superficially resemble swifts, but the families not closely related. Wingbeats less stiff than swifts; perch readily on wires, buildings, trees; migrant roosts in reedbeds and towns can number 1000s. Ages differ, sexes similar or different; adult appearance attained in 1st year in most species.

MARTINS (GENUS *PROGNE*)   New World genus of large swallows with forked tails; often nest in small colonies in cavities in dead trees, cliffs, buildings. Flight powerful, with strong wingbeats, frequent gliding and soaring; often range more widely and higher than other swallows.

GRAY-BREASTED MARTIN *Progne chalybea* 16.5–18cm. Widespread summer migrant in open and semi-open country, in towns, forest clearings, often near water. Fall migrant flocks in Jul–Aug can number in 100s, and mixes readily with other swallows. Both sexes resemble female Purple Martin but slightly smaller with less strongly forked tail, cleaner white underparts (fine dusky streaks rarely visible in field); lacks pale hindcollar but forehead can be paler. Adult male blue-black above, female has dull blue gloss above; juv. dark sooty brownish above, paler throat and chest contrast less strongly with belly than adult. Cf. smaller rough-winged swallows. **SOUNDS:** Chirps slightly higher, burrier, more twittering than Purple Martin, may equally suggest Tree Swallow. Song a chirping, slightly jerky, slow-paced warble. **STATUS:** Common Jan to Sep (a few slightly earlier, and stragglers into early Oct); scare migrant on cayes. (Mexico to S America.)

PURPLE MARTIN *Progne subis* 18–19.5cm. Largest swallow in Belize, a common transient over varied habitats, seasonally in flocks of 100s. Often perches on wires and associates with other swallows. Adult male distinctive, wholly dark, blue-black. Female/imm. variable: note paler forehead and pale hindcollar, deeper tail fork than Gray-breasted; adult female has dark centers to undertail coverts, distinct dark streaks below; juv. whiter below than female, with only faint dark streaks; juv. male bluish above, juv. female brownish above, more similar to Gray-breasted Martin; note deeper tail fork, paler hindcollar of Purple. 1st-summer male resembles female but with variable blue-black feathers on breast. **SOUNDS:** Downslurred, relatively low-pitched, rich twangy *chreu*; higher, slightly burry nasal *chrrih*. **STATUS:** Fairly common to common mid-Jun to Oct (stragglers into Nov), and Feb–Apr (a few from late Jan and into early May). (Breeds N America to Mexico, winters mainly S America.)

RICHMOND'S
SWIFT

CHIMNEY SWIFT

GRAY-BREASTED MARTIN

female

male

PURPLE MARTIN

juv.

1st-summer
male

female

male

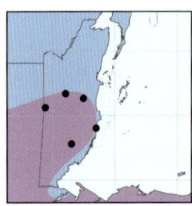

NORTHERN ROUGH-WINGED SWALLOW *Stelgidopteryx serripennis* 12–13.5cm. Fairly chunky, rather plain swallow of open and semi-open country, forest edge, often near water; nests in holes in banks, road cuts, buildings. Tail slightly notched, appears rounded when spread. Wingbeats smooth and floppy, not snappy; rarely soars and glides for prolonged periods, cf. martins. Note brown upperparts, dingy brownish breast (some breeding birds have buffy throat), notched tail, white undertail coverts (can have 1–2 dark subterminal spots). Cf. Ridgway's Rough-winged Swallow. SOUNDS: Slightly wet, buzzy *zzurt* and *zzrih*; harsher calls when alarmed, often in rapid series. STATUS: Breeding distribution needs clarification vs. Ridgway's Rough-winged Swallow (known Belize breeding sites shown with black dots). Local breeder at scattered locations in south. Fairly common Oct–Apr; more widespread in migration, mainly late Jul–Oct, Mar–May. (Breeds N America to Costa Rica, winters Mexico to Panama.)

*RIDGWAY'S ROUGH-WINGED SWALLOW *Stelgidopteryx [serripennis] ridgwayi* 12.5–14cm. Fairly chunky, rather dark swallow of open and semi-open country, forest edge and clearings, mainly in limestone areas; nests in caves, sinkholes, Maya ruins. Rarely in flocks of more than 50 birds. Slightly larger than Northern Rough-winged Swallow, with more strongly notched tail; note darker plumage, black distal undertail coverts, pale lore spots, subtly different voice. Also cf. Gray-breasted Martin. SOUNDS: Hard buzzy *zzzrrt* and *zzrih*, slightly lower, harsher, and more rasping than Northern Rough-winged; spluttering short warbles. STATUS: Fairly common locally, mainly in limestone foothills (known breeding sites shown with black dots). (Mexico to Belize.)

(AMERICAN) CLIFF SWALLOW *Petrochelidon pyrrhonota* 12–13.5cm. Widespread transient migrant in open and semi-open areas, towns, villages, often near water. Associates with other swallows, readily perching on wires. Flight less smooth and graceful than Barn and rough-winged swallows, wingbeats rather choppy. From Cave Swallow by contrasting dark throat; whitish forehead of n. migrants distinctive (migrants from Mexican breeding population typically have rusty forehead); rump often paler than Cave Swallow; wing molt in winter, s. of Belize, vs. late summer–early winter in Cave Swallow. Juv. has weaker head pattern, browner back, like adult by spring. SOUNDS: Burry *chrreh* and variations; alarm call a downslurred, twangy nasal *chiehr*. STATUS: Fairly common to common, especially in fall; migration mainly mid-Aug to mid-Oct (smaller numbers from late Jul and stragglers reported into Nov, but cf. Cave Swallow), late Mar to early May (a few from Feb and into late May). (Breeds N America to Mexico, winters S America.)

*CAVE SWALLOW *Petrochelidon fulva* 11.5–13.5cm. Scarce migrant; any 'Cliff Swallow' later than October should be double-checked. Habits much like Cliff Swallow and the two species may associate together. Note rusty cheeks and pale rusty throat, contrasting poorly with pale underparts, cf. Cliff Swallow. Juv. duller overall, with ghosting of adult pattern. Comprises 2 groups in Belize that may represent species: **Northern Cave Swallow** *P. [f.] pallida* averages larger and paler than **Yucatan Cave Swallow** *P. [f.] citata*, but often difficult to distinguish the groups under typical field conditions. SOUNDS: **Northern**: upslurred *zwieh*, suggests Barn Swallow; alarm a nasal, downslurred *chieh*. **Yucatan**: calls average richer. STATUS: **Northern** is a scarce fall transient Sep–early Dec in south, mainly on s. coast. **Yucatan** is a vagrant (Feb–Mar) in n. Orange Walk. (Breeds Mexico and sw. US to Greater Antilles, winters to Cen America.)

NORTHERN ROUGH-WINGED SWALLOW

juv.

Northern migrant

Resident

RIDGWAY'S ROUGH-WINGED SWALLOW

CLIFF SWALLOW

Mexican

juv. (Jul–Nov)

Northern

CAVE SWALLOW

Yucatan

juv. (Jul–Nov)

Northern

MANGROVE SWALLOW *Tachycineta albilinea* 11–12cm. Small spritely swallow usually near water, from mangroves and coastal lagoons to rivers well inland; nests in cavities in dead trees, limestone rocks, buildings. Flight fast and twinkling, often low over water; perches low on sticks, rocks in water, as well as with other swallows up on wires. White rump diagnostic; also note white underwing coverts (underwings dusky overall on Tree Swallow), small white forehead chevron. Juv. dusky gray-brown above and dingier white below, attains adult plumage by winter. **SOUNDS:** High chipping *chrrit* and *chiri-chrit*; burrier than Tree Swallow. Song a varied series of chirps and burry chips. **STATUS:** Fairly common to common locally, including Ambergris and Caye Caulker. (Mexico to Panama.)

TREE SWALLOW *Tachycineta bicolor* 13.5–14.5cm. Rather chunky migrant swallow, usually near water, especially coastal wetlands. Roosts can number 1000s, usually in extensive wetlands with reedbeds. Flight powerful, often direct, recalling martins rather than smaller, more 'twinkling' Mangrove Swallow. Note chunky build, notched tail, white underparts, dusky underwings. 1st-year female brownish above with variable dusky breast band; adult female averages duller than male, slightly dingier white below. **SOUNDS:** Chirping *chrit* and *chri-chii*, lower, more liquid and gurgling than Mangrove Swallow. **STATUS:** Widespread, nomadic, and locally common Nov–Mar, with small numbers occasional from mid-Sep and into Apr. (Breeds N America, winters s. US to Panama.)

BANK SWALLOW (SAND MARTIN) *Riparia riparia* 11.5–12.5cm. Small, rather compact migrant swallow usually near water, often with other swallows. Often flies fairly low over water, perches on wires. Distinctive, with cleft tail, white throat and neck sides offset by brown breast band. Juv. has wing coverts and tertials edged cinnamon. Cf. rough-winged swallows. **SOUNDS:** Rolled gravelly *zzzr*, often doubled, and buzzy twittering, drier and buzzier than rough-winged swallows. **STATUS:** Fairly common transient, mainly mid-Aug to early Nov, Apr to mid-May; occasional reports in winter. (Breeds Holarctic; winters Mexico to S America.)

BARN SWALLOW *Hirundo rustica* 12.5–14.5cm + streamers. Distinctive, slender, fork-tailed migrant swallow widespread in open and semi-open habitats, often near human habitation and water. Roosts can number 1000s, sometimes on wires in towns. Adult has dark rusty throat, diagnostic long tail streamers (averaging longer on male), variable rusty to buff wash below (can fade to whitish in winter). Juv. has shorter tail, throat and forehead can fade to whitish; note white subterminal tail spots. Winter birds often in wing molt. **SOUNDS:** Upslurred squeaky *jit* and *zwieh*, often run into twittering series; agitated call a clipped *pi-chip*. **STATUS:** Common nonbr. migrant; migration mainly mid-Jul to Nov, mid-Mar to mid-May (rarely into early Jun); sporadic and local in winter. (Breeds Holarctic; winters Mexico to S America.)

Mangrove Swallow

juv.
(Jun–Sep)

adult

Tree Swallow

1st-year
female

male

Bank Swallow

Barn Swallow

faded juv.

fresh adult

OVENBIRDS (FURNARIIDAE; 15 SPECIES) Large Neotropical family most diverse in South America. Plumage mainly shades of brown, many species with a pale wingstripe visible in flight. Ages/sexes mostly similar. Voice and behavior often helpful for ID.

WOODCREEPERS (9 species). Formerly considered a separate family and, as name suggests, typically creep and hitch on trunks and branches, like woodpeckers. For ID note overall size, habitat, behavior, bill size and shape, extent of any paler streaks and spots, voice (sing mostly very early and at dusk).

IVORY-BILLED WOODCREEPER *Xiphorhynchus flavigaster* 22.5–25cm. The most vocal, conspicuous, and widespread woodcreeper in Belize, found in most forest types, adjacent second growth, mangroves. Forages on trunks and larger branches, in bromeliads, often rather sluggish; regular at army ant swarms. Note relatively large size, long stout bill (not really ivory-colored, often dusky pinkish overall, sometimes with mostly dark maxilla), bold pale back streaks; voice. Cf. Streak-headed Woodcreeper. **SOUNDS:** Common call a fairly abrupt slurred whistle, *tcheu.* Song distinctive: an overall descending, laughing series of clear whistles, often slowing and slurring slightly at end; at times starts hesitantly, at other times followed by an abrupt, upslurred *whee whee-wheep!* that is also given separately; songs mostly 2–6 secs, sometimes longer series, rising and falling. **STATUS:** Fairly common to common. (Mexico to nw. Costa Rica.)

STREAK-HEADED WOODCREEPER *Lepidocolaptes souleyetii* 19–20.5cm. Medium-size, slender-billed woodcreeper of open forest and edge, second growth with taller trees, hedgerows. Mainly at mid–upper levels on trunks and branches; often fairly quick and active. Note fine, mostly pale bill, and weakly streaked back. Cf. Ivory-billed Woodcreeper. **SOUNDS:** Song a rapid rolled trill or rattle, 1.5–2 secs; suggests Grayish Woodcreeper but lower, less liquid, slower-paced. Call a short trilled *eeihrrr* or *chirrr,* with similar quality. **STATUS:** Fairly common to uncommon. (Mexico to nw. S America.)

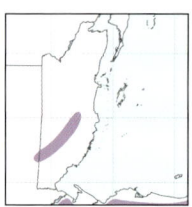

*NORTHERN SPOTTED WOODCREEPER *Xiphorhynchus erythropygius* 23–24cm. Fairly large woodcreeper of cloud forest and humid pine-evergreen forest in Maya Mts. Forages low to high on trunks and larger branches, often at bromeliads. Distinctive, with pale buff goggles, spotted back and underparts; fairly stout straight bill extensively dark above. Cf. Ivory-billed Woodcreeper. **SOUNDS:** Downslurred whistled *whieu.* Song a descending, unhurried series of 2–4 rich to slightly plaintive, downslurred whistles, *wheeo, wheeo, wheeo,* about 1/sec. **STATUS:** Fairly common at higher elevations in Maya Mts., mainly above 700m. (Mexico to nw. Nicaragua.)

STRONG-BILLED WOODCREEPER *Xiphocolaptes promeropirhynchus* 30–31.5cm. Very large woodcreeper of humid forest, pine forest. Feeds low to high, often in bromeliads; joins mixed flocks of larger birds. Note very stout, grayish bill, whitish throat often bordered by broad dark mustache, overall plain back. **SOUNDS:** Song a loud, often slightly descending series of paired whistles, with jerky, ratcheting cadence, mostly 6–10 secs: higher 1st part slurred, 2nd part shorter, fairly abrupt, *chooh'ih chooh'ih.....* Calls infrequently, a muffled, drawn-out snarl slurred into a short emphatic cluck, *ryehhr chk!* **STATUS:** Uncommon to rare and local. (Mexico to S America.)

NORTHERN BARRED WOODCREEPER *Dendrocolaptes sanctithomae* 26–28cm. Distinctive large woodcreeper of humid forest; often at army ant swarms, where sits quietly and is overlooked easily. Barring can be difficult to see in shady forest, and often looks rather dark and plain; note dark lores, stout blackish bill with pale pinkish base. **SOUNDS:** Song a slightly ascending and intensifying series of upslurred, 2-syllable, twangy whistles, each ending with a sharp upward inflection, *duwih' duwih' duwih'...,* mostly 4–6 secs, with distinctive pulsating cadence. Calls infrequently, a quiet *wh-whee* while foraging. **STATUS:** Uncommon to fairly common. (Mexico to nw. Ecuador.)

IVORY-BILLED
WOODCREEPER

STREAK-HEADED
WOODCREEPER

NORTHERN SPOTTED
WOODCREEPER

STRONG-BILLED
WOODCREEPER

NORTHERN BARRED
WOODCREEPER

*PLAIN XENOPS *Xenops [minutus] genibarbis* 11–12.5cm. Small arboreal ovenbird of humid forest, adjacent taller second growth, gallery forest. Forages mainly at mid-levels in fairly open subcanopy on twigs and smaller branches, among vine tangles; rather agile, climbing and often hanging upside-down like a chickadee, hammering at twigs; does not use its tail for support. Ones and twos often with mixed flocks. Distinctive, with contrasting white whisker, wedge-shaped bill, habits. SOUNDS: Song a high, fast-paced, rippling trill, overslurred and slowing at the end, 1–2 secs; typically starts with 1 or more high *tsip* or *pip* notes; suggests Grayish Woodcreeper but higher, thinner, more rippling. Calls a high thin *tseep* and hissing *psssi*. STATUS: Fairly common to uncommon. (Mexico to S America.)

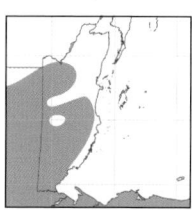

*NORTHERN WEDGE-BILLED WOODCREEPER *Glyphorynchus [spirurus] pectoralis* 14–15cm. Very small, short-billed woodcreeper of humid forest, taller second growth. Forages low to high on trunks of trees all sizes; often fairly active. Joins mixed flocks. Often looks rather dark: note small size, wedge-shaped bill, pale eyebrow, spotted chest. SOUNDS: Song a high, squeaky, twittering crescendo, overall slightly ascending, ends abruptly, 1–2.5 secs. Sharp, high chipping *chrrik*, often doubled, and longer series of chips that may suggest a scolding squirrel. STATUS: Fairly common in south, uncommon to fairly common and more local in north. (Mexico to S America.)

*GRAYISH [OLIVACEOUS] WOODCREEPER *Sittasomus [griseicapillus] griseus* 15–16cm. Small plain woodcreeper of humid forest, taller second growth, gallery forest. Note small size, lack of streaking, small slender bill. Forages low to high on trunks of trees all sizes; often fairly active, spirals up one tree before dropping to base of nearby tree and starting again. Joins mixed flocks. SOUNDS: Song a fast-paced, overslurred, liquid trill, usually about 1 sec; cf. Plain Xenops. Also quiet churring trills that can last 2–3 mins, and a short, dry, rattling trill that suggests Northern Gnatwren. STATUS: Fairly common. (Mexico to n. S America.)

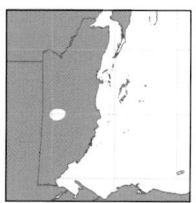

TAWNY-WINGED WOODCREEPER *Dendrocincla anabatina* 18–19cm. Rather chunky woodcreeper of humid forest, taller second growth. Mainly at low to mid-levels, often on thin trunks at ant swarms when several birds may gather; can remain still for long periods and is overlooked easily. Note pale eyebrow, strongly bicolored wings, whitish throat; often raises slightly bushy crest. SOUNDS: Plaintive, sometimes sharp slurred whistle, *tcheu!* or *tchee-u*, at times repeated steadily. Song (?) an insistent, prolonged staccato rattle of rough nasal chips almost slow enough to count, *chri-chri…*, sometimes to over a min. STATUS: Fairly common (at ant swarms, otherwise seen infrequently) to scarce. (Mexico to w. Panama.)

RUDDY WOODCREEPER *Dendrocincla homochroa* 19–20cm. Rather chunky woodcreeper of humid forest, adjacent second growth. Mainly at low to mid-levels, often on thin trunks at ant swarms when several birds may gather; can remain still for long periods and is overlooked easily. Note overall bright ruddy plumage, bushy face with paler eyering. SOUNDS: High, reedy, slightly plaintive slurred *sreeah* and more drawn-out *tleeeoo*. Song a slightly descending, churring rattle, 3–4 secs. Also a prolonged, churring rattle of harsh wooden chips, up to a min or longer, slowing slightly at end; faster-paced than Tawny-winged, notes too fast to count. STATUS: Fairly common (at ant swarms, otherwise seen infrequently) to scarce. (Mexico to nw. S America.)

PLAIN XENOPS

NORTHERN WEDGE-BILLED
WOODCREEPER

GRAYISH
WOODCREEPER

TAWNY-WINGED
WOODCREEPER

RUDDY
WOODCREEPER

RUFOUS-BREASTED SPINETAIL *Synallaxis erythrothorax* 14.5–16.5cm. Small, long-tailed bird of dense second growth, overgrown shrubby clearings, river edge thickets, not in forest. Distinctive, but skulking and rarely seen; mainly detected by voice. Note grayish head, bright rusty breast merging into bright rusty wings, variable black throat patch. SOUNDS: Most often heard is a sneezy *witchew!* Initial note typically given a few times, *whí-whí-witchew*, sometimes repeated steadily. Also a hard *cheurr* and nasal, slightly barking *kyow*. Excited pairs duet with rapid rolled chattering. STATUS: Fairly common locally in south, uncommon and more local in north. (Mexico to nw. Honduras.)

FOLIAGE-GLEANERS (2 species). Medium-size ovenbirds with buff spectacles, rusty tails; often best detected by voice. The 2 species separate by elevation in Belize, with Scaly-throated very local and unlikely to be seen without mounting an expedition. Nest in burrows in banks.

SCALY-THROATED FOLIAGE-GLEANER *Anabacerthia variegaticeps* 16–17.5cm. Remote highland cloud forest in south. Forages acrobatically, low to high in trees and bushes, probing in mossy tangles and epiphytes. Often in pairs or small groups, regularly with mixed flocks. Note bold spectacles, grayish crown, habits; cf. larger Middle American Foliage-gleaner (mainly lowlands). SOUNDS: Common call a slightly harsh, emphatic *skweer!* or *squeezk!* Song a steady to accelerating series of sharp, high, slightly metallic to squeaky chips, mostly 5–11 secs, at times intensifying toward the end and including paired notes, *chiih! chiih! chiih!....* STATUS: Known only from higher elevations on Doyle's Delight, where fairly common; may also occur elsewhere in Maya Mts. (Mexico to w. Ecuador.)

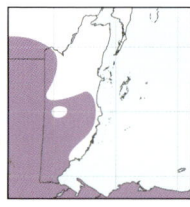

*MIDDLE AMERICAN [BUFF-THROATED] FOLIAGE-GLEANER *Automolus [ochrolaemus] cervinigularis* 19–20.5cm. Understory of humid forest, rarely at edges. Skulks mainly at low to mid-levels in shady understory, probing in dead-leaf clusters; also digs in leaf litter, and ranges rarely to subcanopy when with mixed flocks. Sings from low perch, shivering its tail with each song. Note bold buff spectacles, rich buffy throat, stout bill, habits. SOUNDS: Low gruff *chuk*, sharp nasal *pe-duk*, and hard scolding *tchehrr*. Song a downslurred, chortling rattle, often repeated steadily early and late in the day; 1–1.5 secs. STATUS: Uncommon to fairly common. (Mexico to nw. Panama.)

LEAFTOSSERS (GENUS *SCLERURUS*) (2 species). Stocky, mostly terrestrial ovenbirds with rather long slender bills, relatively short tails. Most often detected by voice, especially call notes as birds flush from forest floor with a quiet whirr of wings, when they may perch on a low branch before dropping back to the ground and eponymously tossing leaves. Nest in burrows in banks.

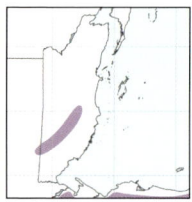

TAWNY-THROATED (MIDDLE AMERICAN) LEAFTOSSER *Sclerurus mexicanus* 16–17cm. Remote highland cloud forest in south. Favors shady forest floor, especially tangled gullies, overgrown banks. Forages in leaf litter; sings from ground and low perches. Typically gives sharp alarm call when flushed from forest floor. Nothing very similar in Belize, but rarely seen in its limited range. SOUNDS: Sharp explosive *sweek!* Song a slightly descending series of 2–7 high, thin, overslurred whistles, *Squeeih squeeih, squeeih*, sometimes run into a short rippling trill; mostly 2–3 secs. STATUS: Uncommon at higher elevations in Maya Mts., mainly above 900m. (Mexico to Panama.)

SCALY-THROATED LEAFTOSSER *Sclerurus guatemalensis* 17–18cm. Chunky dark bird of shady floor of lowland rainforest, often in rather open areas with abundant leaf litter. Forages by tossing leaves with its bill; sings from ground and low perches. Distinctive, but rarely seen unless flushed, when gives sharp alarm call. SOUNDS: Sharp explosive *sweeik!* Song a fairly rapid, slowing and speeding series of bright whistles with characteristic rippling cadence, at times ending with a short twittering trill, 2–5 secs; when excited, songs repeated steadily with barely a pause between them. STATUS: Uncommon to fairly common. (Mexico to nw. Ecuador.)

Rufous-breasted Spinetail

juv.

Scaly-throated Foliage-gleaner

Middle American
Foliage-gleaner

Tawny-throated
Leaftosser

Scaly-throated
Leaftosser

TYPICAL ANTBIRDS (THAMNOPHILIDAE) (9 species). Large Neotropical family most
diverse in South America. Plumage predominantly shades of black, white, grays, and browns. Ages/sexes different or similar; adult appearance attained in 1st year. Voice and behavior helpful for ID.

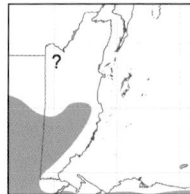

RUSSET ANTSHRIKE *Thamnistes anabatinus* 14.5–16cm. Sluggish and easily overlooked arboreal antbird of humid forest. Mainly at mid–upper levels, foraging in vine tangles, dead-leaf clusters; often joins mixed flocks. Distinctive but uneventful, with stout hooked bill, broad pale eyebrow; sexes similar, but male has concealed tawny patch on back, flared in interactions. **SOUNDS:** High, thin, slightly sibilant *tsip-si*, and upslurred *sweek*. Song a high slurred whistle followed by unhurried series of (usually 4–6) high whistled notes, *tssieur, tsiu-tsiu-tsiu-tsiu-tsiu*, 2–2.5 secs; similar to some songs of Green-backed Sparrow. **STATUS:** Fairly common at higher elevations in Maya Mts., uncommon to scarce in foothills. (Mexico to n. Peru.)

BLACK-CROWNED [SLATY] ANTSHRIKE *Thamnophilus atrinucha* 14–15cm. Medium-size antshrike of lowland forest understory and edge, second growth. Fairly skulking, mainly at low to mid-levels in thickets and tangles; usually in pairs that sometimes associate with mixed flocks. Male distinctive, with bold white spots on wings and tail, black cap, stout bill; white back patch usually concealed, flared in interactions. Female usually with male; note bold wing and tail spotting. Imm. male resembles adult male but has brownish wings. **SOUNDS:** Song (given by both sexes) a rapid series of (about 15–30) nasal *cah* or *aah* notes ending with a more emphatic *ahk*, 1.5–2.5 secs; suggests Barred Antshrike but more even-paced, averages slightly lower, often ends less emphatically. Calls include a nasal cawing *aáanh*, overall descending and sometimes doubled; and a low purring growl, *ah'rrrrrrrrr*, about 1.5 secs. **STATUS:** Fairly common locally. (Belize to w. Ecuador.)

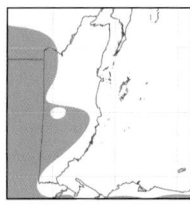

PLAIN ANTVIREO *Dysithamnus mentalis* 11–12.5cm. Small chunky antbird of rainforest; arboreal. Often in pairs, foraging methodically at low to mid-levels in open understory; joins mixed flocks. Distinctive, with heavy bill, shortish tail, narrow pale wingbars; note dark cheeks of male, rusty cap and white eyering of female; cf. Tawny-crowned Greenlet (p. 226), sometimes in the same flocks. **SOUNDS:** Low nasal *nyeu-nyeut*; nasal barking *kah*, often repeated fairly steadily (suggests quiet Barred Forest Falcon). Song an accelerating, laughing, and overall descending series of hollow, slightly nasal whistles, *hyu-hyu-hyu-hyuhyu…*, mostly 2–3 secs; recalls Barred Antshrike but higher, more laughing, lacks final snarl. **STATUS:** Uncommon to locally fairly common. (Mexico to S America.)

DOT-WINGED ANTWREN *Microrhopias quixensis* 11–12cm. Small, attractive, and distinctive little bird of rainforest and edge, especially leafy tangles. Usually in pairs or small groups, moving actively in leafy foliage at low to mid-levels in understory and at edges; often independent of mixed flocks. Note bold white wing spots, long graduated tail tipped white. **SOUNDS:** Song a rapid bouncing-ball series of high, thin, slightly squeaky chips, *pii, pii, pii-pii-pii…*; mostly 2–3 secs; cadence can suggest Dusky Antbird, but notes much higher, thinner. Varied calls mostly chipping and piping whistles, including loud clear *tchip teeoo*, sharp liquid *tew*, and a harsh mew. **STATUS:** Fairly common to common. (Mexico to S America.)

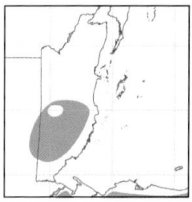

SLATY ANTWREN *Myrmotherula schisticolor* 10–11cm. Scarce in humid foothill forest; typically forages in pairs with mixed flocks at mid-levels in understory. Note small size, rather short tail; male distinctive, female notably plain, with rather blank face, ochre-buff underparts. Cf. larger, long-tailed Dusky Antbird, which differs in behavior and voice. **SOUNDS:** Short, nasal, downslurred mewing *meah* or *nyieh*; infrequently heard song a short, measured series of (usually 2–9) rising, slightly nasal whistles, *wiep, wiep…*, about 2/sec. **STATUS:** Uncommon at higher elevations in Maya Mts., scarce to rare in foothills. (Mexico to S America.)

RUSSET ANTSHRIKE

female

BLACK-CROWNED ANTSHRIKE

male

PLAIN ANTVIREO

male

female

DOT-WINGED ANTWREN

male

female

imm.
male

female

male

SLATY ANTWREN

BARRED ANTSHRIKE *Thamnophilus doliatus* 16.5–18cm. Most widespread and familiar antbird, a poster child for the family. Found in second growth, forest edge, open woodland, thickets; rarely in forest interior. Mostly skulking but still seen fairly frequently, usually in pairs foraging methodically at low to mid-levels in tangles. Male unmistakable; female might suggest a wren but note staring pale eyes, stout hooked bill, spiky erectile crest, unbarred wings and tail. SOUNDS: Song distinctive (given year-round by both sexes), a fairly rapid, laughing series of nasal notes, accelerating toward the end before ending abruptly with an emphatic yap, *ah-ah-ah…anh!*; 2–3 secs. Common call a low, slurred growl, *ahrrrr*. STATUS: Fairly common to common on mainland, uncommon to scarce on Ambergris Caye. (Mexico to S America.)

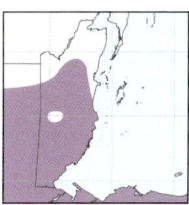

GREAT ANTSHRIKE *Taraba major* 19–21cm. Easily heard but rarely seen, a skulking denizen of second-growth thickets, leafy vine tangles, and humid forest edge. Usually in pairs, foraging at low to mid-levels in shady tangles. Distinctive when seen, with very heavy hooked bill, staring red eyes, clean white underparts. SOUNDS: Song (given by both sexes) a bouncing-ball series of hollow whistles accelerating toward the end and fading quickly; often ends with a short quiet snarl, hard to hear at a distance, *koh, koh…*, 3–4 secs; cf. song of Black-headed Trogon. Hard *chak* notes and growls, and a gruff rattle. STATUS: Uncommon to fairly common. (Mexico to S America.)

DUSKY ANTBIRD *Cercomacroides tyrannina* 12.5–14cm. Notably skulking, heard far more often than seen. Dense second growth and humid forest-edge tangles; rarely wanders into shady forest understory. Usually in pairs, moving low and furtively in thickets; not with mixed flocks. Distinctive, if plain; note narrow whitish wingbars of male, rather blank face and pale rusty underparts of female. SOUNDS: Song a rapid series of usually bright piping whistles, accelerating and then fading abruptly, *pyi-pyi…*, 2–2.5 secs; at times pairs duet, female(?) gives shorter, higher, slightly rising song. Calls include a harsh rasping *brreeea* (cf. White-collared Manakin) and hard churring *kehrrr*. STATUS: Fairly common to common. (Mexico to S America.)

BARE-CROWNED ANTBIRD *Gymnocichla nudiceps* 15–16cm. Scarce but distinctive, medium-size antbird of humid forest edge and second growth, especially swampy *Heliconia* thickets. Usually in pairs on or near ground, often attending army ant swarms. Note stocky build, bright blue facial skin (and crown on adult male); male has distinct white wingbars, female rusty brown overall with paler wingbars. SOUNDS: Song a chant of (usually 8–16) downslurred, slightly nasal whistles, *tcheu tcheu…*, often accelerates slightly at the end, 3–4 notes/sec. Calls include a harsh mewing *meéahr*, and an abrupt *sweik!* suggesting a leaftosser. STATUS: Scarce to uncommon and local. (Belize to n. Colombia.)

ANTTHRUSHES (FORMICARIIDAE; 1 SPECIES) Small Neotropical family of terrestrial, rather crake-like antbirds. Ages differ slightly; weak juv. plumage soon replaced; sexes similar.

MAYAN (MEXICAN) ANTTHRUSH *Formicarius moniliger* 18–19.5cm. Heard much more often than seen. Distinctive, rather chunky and long-legged; walks like a crake on the floor of humid forest, tossing leaves aside with its bill. Sings from ground while walking, also from low perches on vines. Flushes strongly, usually accompanied by sharp alarm call. SOUNDS: Song distinctive, carries well: a rapid, ringing series of about 10 sharp mellow whistles preceded by a single, sharper note: *piu, piu-piu-piu…*, 2–2.5 secs. Call a clipped, hollow *p'tuk!* or *ch-wk*, at times run into rapid clucking series. STATUS: Fairly common to uncommon. (Mexico to Honduras.)

female

male

BARRED ANTSHRIKE

female

GREAT ANTSHRIKE

male

DUSKY ANTBIRD

female

male

BARE-CROWNED ANTBIRD

female

MAYAN ANTTHRUSH

male

imm. male

BECARDS, TITYRAS, AND ALLIES (TITYRIDAE; 8 SPECIES) Recently recognized Neotropical family; several species formerly treated as cotingas, and also includes Northern and Speckled Mourners. Ages/sexes differ strikingly or similar; attain adult appearance in 1st year.

BECARDS (GENUS *PACHYRAMPHUS*) (4 species). Chunky, rather large-headed, and often sluggish arboreal birds, easily overlooked if not vocal. Usually found as singles and pairs, regularly at fruiting trees. Ages/sexes usually differ; 1st-year male plumage variably intermediate between adult male and female.

ROSE-THROATED BECARD *Pachyramphus aglaiae* 16–17cm. Most widespread and familiar becard, with big bushy head, ungraduated tail. Found in wide variety of wooded and forested habitats, from mangroves to pine forests. Forages low to high, often at fruiting trees with tityras; ranges into weedy fields, perching on fence lines; joins mixed flocks. Male distinctive, but rose throat can be lacking in south. Female told from smaller Gray-collared and Cinnamon Becards by bushy blackish cap without distinct pale brow, ungraduated tail; also note plain wings cf. Gray-collared. 1st-year male variably intermediate between adult male and female, often with some pink on throat. **SOUNDS:** Plaintive downslurred *tseeu*, often run into a rolling and spluttering reedy chatter or trill, or a squeaky chatter may slur into a downslurred *tew*. Dawn song a high, slightly reedy, plaintive *si-tchew, wii-chew* or simply *si-tchew*, repeated every 2–6 secs. **STATUS:** Fairly common in north, uncommon to scarce and local in south; occasional on Ambergris Caye. (Mexico and sw. US to w. Panama.)

GRAY-COLLARED BECARD *Pachyramphus major* 14–15.5cm. Attractive, widespread becard, nowhere common. Singles and pairs occur in wide variety of wooded and forested habitats. Mainly at mid–upper levels, often with mixed flocks. Male distinctive, with pale gray underparts and brow; back variably mottled black, to mostly gray in north; on female note patterned wings, graduated tail, cf. Rose-throated Becard. **SOUNDS:** Song comprises variations on 2 themes: series of (usually 3–20) paired, mellow whistles, *hu-widíh hu-widíh…*, about 1 phrase/sec, 2nd part strongly upslurred; also slightly descending series of (usually 5–7) plaintive whistles, often speeding up slightly at end when may run into a quick short roll. Varied nasal chips and slightly squeaky chippering twitters when agitated. **STATUS:** Scarce to uncommon and local, perhaps wandering more widely in winter. (Mexico to nw. Nicaragua.)

CINNAMON BECARD *Pachyramphus cinnamomeus* 14–15cm. Rather small becard of rainforest edge, adjacent clearings with taller trees. Sexes alike. Usually at mid–upper levels, sometimes with mixed locks. Note subtle face pattern with dark lores, short pale brow, dark rusty cap, cf. female Rose-throated Becard, much larger and plainer Rufous Mourner. **SOUNDS:** Varied series of high reedy whistles, at times rapid twittering series, and a plaintive *seeeeiu*. Song comprises 1–2 drawn-out, plaintive high whistles followed by a faster, slightly descending series of (usually 3–9) shorter notes, *teeeu dee-dee…*, or *cheei deu-deu-deu*. **STATUS:** Uncommon to fairly common. (Mexico to w. Ecuador.)

WHITE-WINGED BECARD *Pachyramphus polychopterus* 14–15cm. Attractive small becard of rainforest edge, adjacent clearings and second growth with taller trees. Mainly at mid–upper levels, often in canopy of fruiting trees. Attractive male distinctive, cf. paler Gray-collared Becard, which has pale brow. Female also distinctive but can be puzzling; note broken pale eyering, cinnamon wing edgings. **SOUNDS:** Song comprises variations on 2 themes: series of (usually 4–9) mellow, downslurred whistles, the 1st typically longer, then speeding and slowing at end, *tcheu chu chu…*, about 6 notes/sec; also 1–2 mellow notes followed by a rapid series of (usually 8–20) descending chips that slow at end, *chiu chi-chi…*, about 10 notes/sec. Varied, twangy whistled phrases when agitated. **STATUS:** Uncommon to fairly common in south; spreading with deforestation, rare and sporadic n. of mapped range. (Mexico to S America.)

ROSE-THROATED
BECARD

north

south

female

males

GRAY-COLLARED
BECARD

male

imm.
male

female

CINNAMON
BECARD

male

female

WHITE-WINGED BECARD

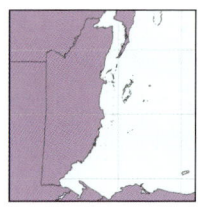

MASKED TITYRA *Tityra semifasciata* 20–22cm. Widespread and often conspicuous in forest and edge habitats, plantations, semi-open areas with fruiting trees, hedgerows. Often in pairs, mainly at mid–upper levels, frequently perching on prominent snags. Flight slightly undulating, suggesting a compact woodpecker. Nests in holes in snags, sometimes stolen from woodpeckers. Distinctive, with naked, reddish-pink face, but cf. superficially similar Black-crowned Tityra. SOUNDS: Short quacking and rasping nasal notes, often doubled or in rhythmic series, *rehk* and *reh-rehk*, or *reh-rehk reh-reh-rehk…*, and variations; could be passed off as a frog. STATUS: Fairly common to common on mainland, uncommon on Ambergris Caye. (Mexico to S America.)

BLACK-CROWNED TITYRA *Tityra inquisitor* 19–20.5cm. Widespread but never numerous in forest and edge habitats, plantations, semi-open areas with fruiting trees. Habits much like commoner Masked Tityra, with which Black-crowned often associates. Distinctive if seen well, cf. Masked Tityra; in flight, note translucent primary panels of Black-crowned, lacking on Masked. SOUNDS: Gruff, slightly rasping *shehk*, singly or in short, at times chattering series that may suggest Mesoamerican [Band-backed] Wren; lacks nasal, burrier quality of Masked Tityra. STATUS: Uncommon to fairly common. (Mexico to S America.)

MANAKINS (PIPRIDAE; 2 SPECIES) Neotropical family of small fruit-eating birds; males often colorful and noted for complex displays. Ages/sexes differ; adult appearance attained in 1–3 years. Tend to sit quietly in shady understory and mid-story, sallying briefly to pluck berries. Easily overlooked if not vocal or displaying.

RED-CAPPED MANAKIN *Ceratopipra mentalis* 9.5–10.5cm. Striking manakin of humid forest, adjacent shady clearings and second growth. Perches low to high, mainly at mid-levels in shady understory, at times with mixed flocks. Male display 'dance' includes various slides, pivots, and leaps along mid-level horizontal branches, accompanied by wing snaps and buzzes. Male unmistakable; female notably drab but note beady eye, pinkish bill and legs; cf. female White-collared Manakin. Imm. male resembles female but eyes pale, often has some red flecks on head (as do some adult females). SOUNDS: Calls include high sharp *píp* and thin, drawn-out whistle, about 1 sec, often followed by sharp chip, *tsssiuu chk!* In display, varied arrangements of sharp, often slightly liquid chips in short series; high overslurred whistles tailing off, often followed by a rasping mechanical buzz or sharp *chk*; plus wing snaps, buzzes, and rattles. STATUS: Fairly common. (Mexico to w. Ecuador.)

WHITE-COLLARED MANAKIN *Manacus candei* 11.5–12cm. Snappy manakin of rainforest edge, adjacent second-growth thickets such as *Heliconia* patches, overgrown plantations. Mainly at low levels in shady thickets, where male often detected by wing whirr during sallies for fruit; less often at mid–upper levels in fruiting trees. Displaying males clear a circle of forest floor and make snapping display leaps between slender saplings either side of the cleared area. Male unmistakable; female told by dark bill, orange legs, yellow belly, cf. female Red-capped Manakin. Imm. male resembles female but throat and median upper breast pale gray. SOUNDS: Downslurred, burry whistled *brréu*, and lower burry *rrreu*. In display, males make remarkably loud 'firecracker' wing snaps and mechanical wing rattles, often alternated with burry whistles; wings make papery rustling in flight. STATUS: Fairly common in south, uncommon and more local in north. (Mexico to w. Panama.)

MASKED TITYRA

female

male

BLACK-CROWNED TITYRA

imm.

female

male

RED-CAPPED MANAKIN

male

female

display

WHITE-COLLARED MANAKIN

male

female

display

NORTHERN [THRUSHLIKE] MOURNER (SCHIFFORNIS) *Schiffornis [turdina] veraepacis* 16.5–17.5cm. Retiring denizen of humid forest; overlooked easily unless singing. Mainly at low levels in fairly open to dense understory; often perches low on thin vertical stalks. Distinctive but rather nondescript: note earth-brown plumage tones, paler eyering, habits, voice. SOUNDS: Haunting whistled song repeated at irregular intervals, 1st note long and overslurred before a sharp upward inflection and abrupt ending, *tjeeeuuu-i wi-chi*, 1.5–2 secs; less often a shorter *tjeeuu wi-chi*. Short low rattle when agitated. STATUS: Uncommon to fairly common. (Mexico to w. Ecuador.)

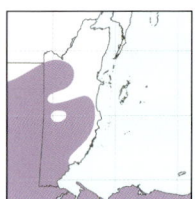

RUFOUS MOURNER *Rhytipterna holerythra* 19.5–21cm. Plain rusty tyrant-flycatcher of rainforest and edge. Mainly at mid–upper levels where sometimes joins mixed flocks; sings from perch in subcanopy. Perches quietly when not singing; flutters after invertebrate prey. Resembles larger Rufous Piha and often occurs in the same areas: note piha's paler eyering and throat, stouter bill with less extensive pale pinkish at base, voice. SOUNDS: Distinctive leisurely wolf whistle, *wheeeeu-heeu*, rising then falling, 1.5–2 secs; also a mournful descending *wheeeu*. Song a steady series of (usually 5–20) plaintive whistles, *wheéu wheéu…*, 10/7–8 secs, at times with a longer introductory *whi'heeeu*; at dawn also *teeuu te-du* repeated. STATUS: Uncommon to fairly common. (Mexico to nw. Ecuador.)

SPECKLED MOURNER *Laniocera rufescens* 20.5–21.5cm. Scarce inhabitant of rainforest; overlooked easily unless singing. Perches upright and still for long periods at mid–upper levels; flutters after invertebrate prey in foliage. Slightly chunkier than Rufous Mourner, with patterned wing coverts, blockier head with pale eyering; also cf. Rufous Piha. Yellow breast tufts flared when singing. SOUNDS: Plaintive, slightly overslurred *wheeeeeu*, about 1 sec, ending fairy abruptly. Song a plaintive, slightly tinny, slurred whistle, *tew-i-tieh*, usually repeated 2–7x in steady series. STATUS: Rare to scarce and local, mainly in south. (Mexico to nw. Ecuador.)

COTINGAS (COTINGIDAE; 2 SPECIES) Neotropical family of fruit-eating birds, occurring mainly in South America. Ages/sexes differ strikingly or similar; like adult in about 1 year.

RUFOUS PIHA *Lipaugus unirufus* 23–26.5cm. Large, plain, rusty cotinga of rainforest; heard more often than seen. Mainly at mid–upper levels; sings from perch in open midstory and subcanopy. Feeds on fruit and invertebrates plucked from foliage. Cf. smaller but remarkably similar Rufous Mourner, which has shallower bill with more extensive pale pinkish at base. SOUNDS: Loud ringing whistles given irregularly, often in response to an abrupt loud noise: *p'wEE-oo!* and *cheEOo! p-wee'oo!* Often simply *pweEOO!* suggesting Pauraque. Hard squirrel-like clucks, often in short 'machine-gun' bursts when agitated. STATUS: Uncommon to fairly common but local, most numerous in Maya Mts. (Mexico to nw. Ecuador.)

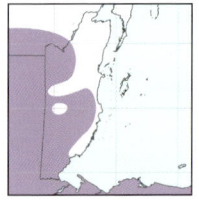

LOVELY COTINGA *Cotinga amabilis* 18–19cm. Stunning and sought-after species of rainforest and edge, adjacent second growth and clearings with taller fruiting trees. Mainly in canopy, where small numbers may gather at fruiting trees; plucks berries in brief sallying flutters. Often sits quietly for long periods, and males especially can perch on exposed snags and be seen from long distance. 'Electric-blue' male unmistakable; female distinctive but can be puzzling: note plump shape with rounded head, spotted underparts, scalloped upperparts. Imm. like female with buffier wing edgings, more spotted vs. scaly upperparts; young males can show patches of adult color. SOUNDS: Abrupt, slightly squeaky high *piic!* easily passed off as a frog. In flight, male makes soft, dry, ticking wing rattle, easily passed over as an insect. STATUS: Scarce to uncommon and local, perhaps nomadic. (Mexico to w. Panama.)

NORTHERN MOURNER

RUFOUS MOURNER

imm.

SPECKLED MOURNER

RUFOUS PIHA

LOVELY COTINGA

female

males

WHISKERED FLYCATCHERS (3 species). Small, enigmatic New World assemblage formerly subsumed within tyrant-flycatchers, now treated as a family (Onychorhynchidae) or merged into Tityridae. Ages/sexes similar overall; all have notably long rictal bristles, or 'whiskers.'

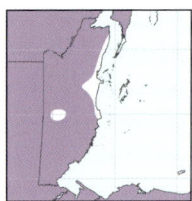

SULPHUR-RUMPED FLYCATCHER *Myiobius sulphureipygius* 11.5–12.5cm. Attractive and active, a distinctive little bird of humid forest. Usually at low to mid-levels in shady understory, often near water; joins mixed flocks. Typically flits with tail fanned and wings drooped to show off pale yellow rump, which often seems to glow in shady forest undersory. Yellow crown patch usually concealed, reduced or absent on female and juv. SOUNDS: Rather quiet. Soft clipped *tlik* given on occasion. STATUS: Fairly common. (Mexico to w. Ecuador.)

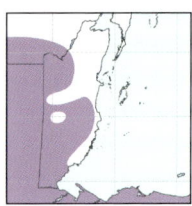

RUDDY-TAILED FLYCATCHER *Terenotriccus erythrurus* 9–9.5cm. Tiny and rather cute inhabitant of rainforest. Singles and pairs perch in fairly open midstory and subcanopy, often on thin vines; sometimes joins mixed flocks. Distinctive, with cinnamon underparts, rusty wings and tail. SOUNDS: Fairly quiet, high, 2-part whistle, 1st note lisping and reedy, 2nd emphatic, *pssii pit*; less often simply *speeu*. STATUS: Uncommon and local. (Mexico to S America.)

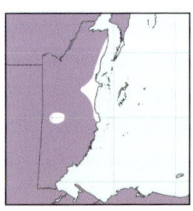

NORTHERN ROYAL FLYCATCHER *Onychorhynchus mexicanus* 16.5–18cm. Medium-size, 'hammerhead' flycatcher of humid forest. Singles or pairs range in mid-levels of shady understory; long scraggly nest hanging over stream or other opening can be good clue to presence. Often active, sallying and fluttering after prey in leafy foliage; joins mixed flocks. Distinctive, with pale cinnamon rump and tail; crest usually held flat, raised mainly in alarm and rarely seen spread open; male crest fiery red, female crest yellow-orange. Juv. upperparts and chest scalloped dusky, soon like adult. SOUNDS: Slightly hollow, plaintive *whee-uk*, suggests a muffled Rufous-tailed Jacamar. Song a descending, slowing series of (usually 5–8) plaintive whistles with short intro note, *whi' peeu, peeu....* STATUS: Uncommon to locally fairly common. (Mexico to nw. Venezuela.)

TYRANT-FLYCATCHERS (TYRANNIDAE; 45+ SPECIES) Large, diverse, and taxonomically vexed New World assemblage of insectivorous and frugivorous birds, ranging from tiny tyrannulets to large and conspicuous kingbirds (also see whiskered flycatchers). Ages usually differ slightly, sexes similar; like adult within 1st year. Many species visually similar, and voice often very important for ID, as is an appreciation of genus characters. Most forest species best detected by voice and can seem rare or even absent until vocalizations are learned.

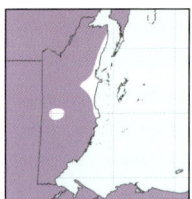

STUB-TAILED SPADEBILL *Platyrinchus cancrominus* 9–10cm. Appropriately named little flycatcher of humid forest. Favors low to mid-levels in shady understory; often sits still for long periods then makes a short sally to pluck food from underside of leaf or twig and moves to new perch. Distinctive, but easily overlooked unless vocal: note compact shape, face pattern. Male has yellow crown patch, usually concealed. SOUNDS: Bright, abrupt nasal *kidik* and *ki-di-dik* reveal presence, repeated irregularly. Dawn song an excited, rolled nasal trill alternated with sharp nasal chips. STATUS: Fairly common. (Mexico to w. Panama.)

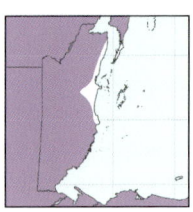

*NORTHERN OCHRE-BELLIED FLYCATCHER *Mionectes [oleagineus] assimilis* 12–13cm. Distinctive but rather plain flycatcher of humid forest and edge, adjacent second growth. Perches low to high, jerking its head back and forth slowly, quickly lifting one wing at a time; often at fruiting trees and shrubs. Sings from perch in fairly open midstory, usually 2–3 males in a loose 'lek.' Note rounded head, slender bill, ochre belly. SOUNDS: Song a repetition of sharply overslurred, bright nasal clucks, every 1–2 secs, alternated with typically slightly faster bursts of downslurred nasal yaps, *whik whik...kyeh-kyeh-kyeh whik whik.....* Infrequent calls include plaintive *cheu* and sharp *plik*. STATUS: Fairly common. (Mexico to n. S America.)

SULPHUR-RUMPED FLYCATCHER

RUDDY-TAILED FLYCATCHER

NORTHERN ROYAL FLYCATCHER

female

nest

males

STUB-TAILED SPADEBILL

NORTHERN OCHRE-BELLIED FLYCATCHER

*MISTLETOE [PALTRY] TYRANNULET *Zimmerius [vilissimus] parvus* 9.5–10.5cm. Inconspicuous small flycatcher of humid forest edge, semi-open areas with scattered trees, hedgerows, open woodland, gardens; especially areas with mistletoe. Forages low to high, regularly in canopy; often perches with tail slightly cocked, plucks berries and snatches insects with short flits and sallies. Best detected (and identified) by call; also note yellow wing edgings (not wingbars), stubby dark bill, whitish eyebrow. Cf. much larger Mesoamerican Elaenia, Sclater's Flatbill. SOUNDS: Overslurred, relatively loud whistled *peeéu* and variations, can be repeated steadily. Dawn song can be repeated steadily, a plaintive whistled *tyeeu chi-ti*, with a flourish suggesting Gray-collared Becard. STATUS: Uncommon in south, mainly in foothills. (Belize to nw. Colombia.)

*MESOAMERICAN [GREENISH] ELAENIA *Myiopagis [viridicata] placens* 13–14cm. Rather slender, long-tailed small flycatcher of humid forest and edge. Perches fairly upright, mainly at mid–upper levels, and often quiet for long periods. Forages with short sallies, plucking insects from foliage, and often eats berries. Distinctive but relatively plain, with short whitish brow, no strong wing markings, faint dusky breast streaking, small narrow bill; yellow crown patch usually concealed. Cf. Sclater's Flatbill. Juv slightly browner overall with dull buffy wingbars, lacks crown patch. SOUNDS: High, slightly reedy downslurred *tseeu* and *seei-seeir*; usually at irregular intervals. Dawn song a high, slightly plaintive slurred *t'cheuuíh*, or *t'cheuuí-úih*, every 2–3 secs. STATUS: Uncommon to fairly common. (Mexico to Honduras.)

*SCLATER'S [YELLOW-OLIVE] FLATBILL (FLYCATCHER) *Tolmomyias [sulphurescens] cinereiceps* 12.5–13.5cm. Rather stocky, large-headed small flycatcher of humid forest and edge, gallery forest. Mainly at mid–upper levels; usually perches with tail slightly cocked. Sallies to snatch insects from foliage; also eats berries. Note pale spectacles, pale eyes, pinkish legs, broad bill with pale mandible, cf. Mesoamerican Elaenia. Juv. has brownish eyes, head variably tinged olive. SOUNDS: Easily passed off as an insect. Song a series of (usually 2–5) high, lisping, slightly shrill upslurred notes in measured, often slightly intensifying series, 1–1.5 secs between notes, such as *ssih sssih sssih sssih*; can be varied to buzzier *zzzieh* at times. Also quiet single *sssi*. STATUS: Fairly common to common. (Mexico to Costa Rica.)

SEPIA-CAPPED FLYCATCHER *Leptopogon amaurocephalus* 12.5–13.5cm. Small, slender-billed flycatcher of humid forest and edge. Mainly at mid-levels in shady understory, at times with mixed flocks; sallies for insects and often eats berries. Frequently flicks open one wing at a time. Note distinctive face pattern with dark cheek patch; also brown crown, pale cinnamon wingbars and tertial edgings. Juv. has brighter, buffier wingbars. SOUNDS: Infrequently heard. Rapid, slightly spluttering, rattled trill, slightly overslurred or descending overall, 1.5–2 secs; at times preceded by 1–2 sharp clucking chips, *whik, whik, prrrrrrrrrrru*. STATUS: Uncommon to fairly common. (Mexico to S America.)

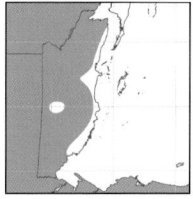

EYE-RINGED FLATBILL *Rhynchocyclus brevirostris* 15–16.5cm. Rather bulky, large-headed flycatcher of humid forest. Mainly at mid-levels in shady understory, where often sluggish and overlooked easily; joins mixed flocks. Nothing very similar in Belize: note bulky shape, contrasting white eyering in grayish face, big bill. SOUNDS: Easily passed off as an insect. Song a series of (usually 2–5) high, lisping, slightly rough upslurred notes, 1st note(s) often longer, such as *zzzzíh zzzi-zzzi-zzi*; notes longer and closer-spaced than more measured song of Sclater's Flatbill. Also single *sshirrrr*, harsher, more steeply rising than Sclater's, and short harsh *zhhih*. STATUS: Uncommon to fairly common. (Mexico to nw. Colombia.)

MISTLETOE TYRANNULET

MESOAMERICAN ELAENIA

SCLATER'S FLATBILL

imm.

SEPIA-CAPPED FLYCATCHER

EYE-RINGED FLATBILL

NORTHERN BEARDLESS TYRANNULET *Camptostoma imberbe* 9.5–10.5cm. Very small drab flycatcher of open forest and edge habitats, semi-open areas with trees and bushes, especially in drier areas. Forages low to high, mainly at mid–upper levels. Posture typically fairly upright with tail below body plane and loosely wagged up and down. Plain but distinctive, with bushy crest, small bill with bright orange base, habitual tail pumping. Fresh plumage has broad cinnamon wingbars, fading to dull pale buff. SOUNDS: Slightly overslurred *peért* and softer, drawn-out *peeeeu*; slightly descending, measured series of (usually 3–8) downslurred notes, *dee dee dee…*; short bubbling *deedl-idl-it*. Song a varied short series of piping reedy whistles, typically including 1–2 louder notes, at times repeated over and over, such as *pi pii pii pee PEE pii-pi*. STATUS: Fairly common in north, uncommon to scarce and local in south. (Mexico to nw. Costa Rica.)

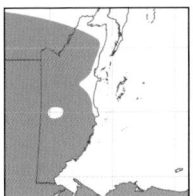

YELLOW-BELLIED TYRANNULET *Ornithion semiflavum* 8–9cm. Tiny, short-tailed flycatcher of rainforest canopy and subcanopy, where overlooked easily. Gleans in foliage, dead-leaf clusters; does not usually sally or hover. Note short tail, yellow underparts, bold white eyebrow. Juv. has duller underparts, brownish cast to upperparts. SOUNDS: Strident, overslurred or rising *sweéh*, at times doubled or in short series; nasal chips and clucks when agitated. Song repeated steadily, a slightly descending series of (usually 3–5) plaintive whistles, *deee-dee-dee-di*, last note sometimes upslurred. STATUS: Uncommon to fairly common. (Mexico to w. Panama.)

COMMON TODY-FLYCATCHER *Todirostrum cinereum* 9–10cm. Handsome little flycatcher of varied open and semi-open habitats, from farmland with hedges and scattered trees to mangroves, forest edge, second growth. Often in pairs, loosely swinging its cocked tail, and darting quickly after prey. Distinctive, with long bill, cocked tail with graduated white tips, staring whitish eyes in blackish hood. Juv. has duller eyes, dingier plumage. SOUNDS: Hard smacking *chk* suggests a warbler, often repeated steadily; short, high twittering trill often repeated quickly 2–6×, *till-ill-ill*, recalls Middle American Kingbird. STATUS: Fairly common to common. (Mexico to S America.)

SLATE-HEADED TODY-FLYCATCHER *Poecilotriccus sylvia* 9–9.5cm. Skulking and easily overlooked in thickets and tangles at humid forest edge, in second growth, riverside bamboo. Hops inconspicuously in dense tangles, rarely in openings and at edges. Told from Northern Bentbill by habits, white spectacles, longer and straight bill. SOUNDS: Abrupt, frog-like cluck and short growling churr, often combined, *pc prrrrr*. STATUS: Uncommon to fairly common but local. (Mexico to S America.)

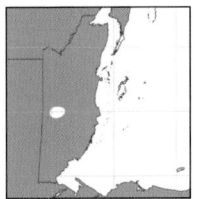

NORTHERN BENTBILL *Oncostoma cinereigulare* 9.5–10.5cm. Inconspicuous inhabitant of humid forest understory and edge. Mainly at low to mid-levels in understory with thin stalks and vine tangles. Perches fairly upright, often leaning forward slightly; sallies in foliage. Tiny size, pale eyes, thin pinkish legs, and eponymous bent bill distinctive; cf. Slate-headed Tody-Flycatcher. Juv. has duller eyes, olive head, buffier wingbars. SOUNDS: Varied low churrs and clucks, most often a drawn-out, frog-like *urrrrrrr*, 1–1.5 secs. Song(?) a slightly stuttering *pirrrip pirrrip p-p-prrrr*, 1st note given 1–3×. STATUS: Fairly common. (Mexico to w. Panama.)

crest can be
held flat

**NORTHERN
BEARDLESS TYRANNULET**

**YELLOW-BELLIED
TYRANNULET**

COMMON TODY-FLYCATCHER

juv.

**SLATE-HEADED
TODY-FLYCATCHER**

**NORTHERN
BENTBILL**

YELLOW-BELLIED ELAENIA *Elaenia flavogaster* 15–16.5cm. Bushy-crested and often conspicuous flycatcher of varied open and semi-open habitats, from overgrown grassy fields to forest edge, often with fruiting shrubs. Mainly gleans from foliage and often eats berries. Note spiky crest, typically raised to show bushy white center; pale pinkish base to mandible; voice. Juv. has shorter crest lacking white base. SOUNDS: Burry, overslurred drawling *breéuh*; hoarse rhythmic bickering *rreeahr-ch'reer...* or *bríka-weehr...*, usually 3–5× in duets. Dawn song a burry *frrí-diyu* or *prri di-di-eu* and variations, over and over every 1–2 secs. STATUS: Fairly common to common, including Ambergris Caye; occasional visitor to inshore cayes. (Mexico to S America.)

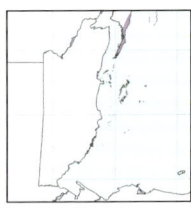

CARIBBEAN ELAENIA *Elaenia martinica* 14–15cm. Notably drab flycatcher of humid forest, thickets, semi-open areas with fruiting trees and shrubs. Perches low to high, often sluggish and inconspicuous. Crest usually held flattened; note bright orange-pink mandible base, voice. Juv. has shorter crest lacking white base. Cf. Yellow-bellied Elaenia, smaller Northern Beardless Tyrannulet. SOUNDS: Slightly emphatic overslurred *péeu*, sometimes burry; downslurred *wheéu*. Dawn song a repetition of short, rich burry phrases every 3–6 secs, such as *frree-bee-byur, frree-bee-byur....* STATUS: Uncommon to fairly common on Ambergris and Caye Caulker; very rare wanderer to other cayes and n. coast mainland. (Mexico, n. Belize, and Caribbean.)

EMPIDS (GENUS *EMPIDONAX*) (6 species).

Small migrant flycatchers with pale wingbars and usually pale eyerings; crown often slightly peaked but not distinctly crested. Unlike pewees, empids do not return repeatedly to same prominent perches and they tend to perch more inconspicuously, several species inside forest or woodland. Habitually flick tails up and twitch wings, unlike pewees. Best identified by voice and structure; also note habitat, overall plumage tones.

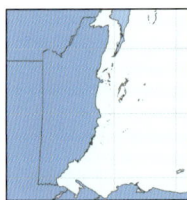

LEAST FLYCATCHER *Empidonax minimus* 11.5–12.5cm. Widespread small empid of open woodland and edge, shrubby second growth; usually at edges and in fairly open situations, not inside forest. See genus note. Mainly at low to mid-levels, often fairly active. Note relatively compact shape, dirty whitish throat, contrasting whitish eyering; relatively blunt and fairly broad bill usually has some dusky below near tip, not uniformly pale. SOUNDS: Sharply upslurred *swik* or *whit*, similar to Willow Flycatcher but slightly higher. STATUS: Fairly common to common Sep–Apr; more widespread in migration, Aug–Oct, Apr–May. (Breeds n. N America, winters Mexico to Costa Rica.)

YELLOW-BELLIED FLYCATCHER *Empidonax flaviventris* 12.5–13.5cm. Widespread empid of humid forest, taller second-growth woodland, plantations. See genus note. Perches low to high, mainly at mid-levels in shady understory and edge. Fairly compact shape recalls Least Flycatcher; note pale yellowish wash to throat and underparts (belly often not especially yellow), voice (but call similar to Acadian Flycatcher); mandible pale orange-pink. Cf. Acadian Flycatcher, a transient in Belize. SOUNDS: Sharply overslurred, slightly explosive *speéik*; more leisurely downslurred *pyeeh*; mainly in spring, a plaintive slurred *tch'wee*, suggesting Eastern Pewee but shorter. STATUS: Fairly common Sep–Apr, especially in south; more widespread in migration, mid-Aug to Oct, Apr–May. (Breeds n. N America, winters Mexico to w. Panama.)

ACADIAN FLYCATCHER *Empidonax virescens* 13–14cm. Relatively large, big-billed, and long-winged transient empid favoring humid forest and edge. See genus note. Perches mainly in shady mid-story and subcanopy, often near water. Note big bill, buff wingbars, dingy whitish throat; call very similar to Yellow-bellied Flycatcher. All ages fresh-plumaged in fall (vs. worn in adults of similar species); also note weaker dark band across base of secondaries than Yellow-bellied. Cf. Alder and Yellow-bellied Flycatchers. SOUNDS: Sharply overslurred, slightly explosive *speéip*, much like Yellow-bellied Flycatcher but averages higher, more explosive. STATUS: Fairly common late Aug–early Nov, uncommon mid-Mar to early May. (Breeds e. N America, winters s. Cen America to nw. S America.)

Yellow-bellied Elaenia

juv.

Caribbean Elaenia

crest rarely raised

Least Flycatcher

imm.

Yellow-bellied Flycatcher

imm.

imm.

Acadian Flycatcher

WILLOW FLYCATCHER *Empidonax traillii* 12.5–14cm. Transient migrant empid of open and semi-open habitats with hedgerows, scattered trees, second growth, marshes; not in forested habitats. See genus note. Note weak eyering, whitish throat, voice. Safely told from Alder Flycatcher by voice; also cf. Least and White-throated Flycatchers, pewees. SOUNDS: Sharply upslurred, mellow *whit* or more liquid *whuit*, similar to Least Flycatcher. STATUS: Fairly common late Aug–Oct, including cayes; uncommon late Apr–early Jun; scarce and local in winter in s. Toledo. (Breeds N America, winters Mexico to S America.)

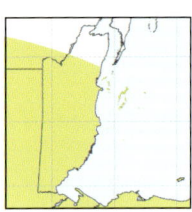

ALDER FLYCATCHER *Empidonax alnorum* 12.5–14cm. Transient migrant empid of open and semi-open habitats with hedgerows, scattered trees, second growth; at times occurs alongside Willow Flycatcher. See genus note. Alder averages brighter, more contrasting than Willow, with more distinct narrow eyering, more sharply defined wingbars and tertial edgings, but only reliably separated by voice. Cf. Least, Acadian, and White-throated Flycatchers, pewees. SOUNDS: High, sharply overslurred *piic* or *peek*, distinct from *whit* of Willow Flycatcher. STATUS: Fairly common late Aug–Oct, including cayes; one confirmed spring record (mid-May), but likely more regular, at least rarely in south; details of status relative to Willow Flycatcher require elucidation. (Breeds n. N America, winters S America.)

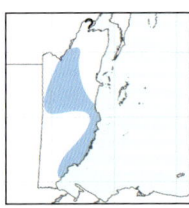

WHITE-THROATED FLYCATCHER *Empidonax albigularis* 12–13.5cm. Rather drab but subtly distinctive empid of marshes with taller vegetation, reedbeds; overgrown damp pastures with bushes, fence lines. See genus note. Typically low and inconspicuous, often perches on fences, weed stalks. Best located and identified by voice. Slightly smaller, more compact than Willow and Alder Flycatchers, with browner upperparts, buff wingbars, ochre tinge to flanks; whitish throat not an especially striking ID feature. SOUNDS: Call quite unlike other empids: relatively low, burry, overslurred *rréah* or *brriéh*. STATUS: Scarce to uncommon and local, mainly Aug–Apr (sometimes from late Jul and into early May). (Mexico to w. Panama.)

PEWEES (GENUS *CONTOPUS*) (5 species). Rather drab flycatchers with slightly peaked or crested heads, paler wingbars; lack distinct pale eyerings. Perch on exposed twigs or prominent high snags and sally out for insects, often returning repeatedly to the same perch and quivering tail upon landing. Unlike empids, pewees do not habitually flick tail and wings.

***NORTHERN TROPICAL PEWEE** *Contopus [cinereus] bogotensis* 13–14.5cm. Inconspicuous small flycatcher of humid forest edge, adjacent second growth and pastures with hedgerows, scattered trees. See genus note. Slightly more compact than Eastern Pewee with shorter primaries, darker crown. Cf. Willow and Alder Flycatchers, best told by habits, voice. SOUNDS: Short, slightly overslurred ringing trill, suggesting Middle American Kingbird; high, sharply overslurred *pyeép*. Song a vaguely disyllabic, high upslurred *pseep* or *p'seíh*, every 1–3 secs; at longer intervals after dawn. STATUS: Fairly common to uncommon. (Mexico to n. South America.)

EASTERN PEWEE (WOOD-PEWEE) *Contopus virens* 13.5–15cm. Transient migrant in wooded and semi-open habitats. See genus note. Longer-winged and slightly larger than Northern Tropical Pewee, with different voice; also cf. Willow and Alder Flycatchers, which have shorter wings, different behavior. Imm. in fall has fresher plumage, buff wingbars. SOUNDS: Plaintive, clear up-slurred *p'weée* ('pewee') or *du-weée*; often vocal in migration. STATUS: Common Aug–Nov (stragglers into Dec), mid-Mar to early Jun; most numerous in fall. (Breeds e. N America, winters S America.)

Very similar **Western Pewee** *Contopus sordidulus* (13.5–15cm) has been recorded in fall from Mountain Pine Ridge, perhaps a scarce (irregular?) transient in s. Belize. Averages duskier and grayer than Eastern Pewee, with more strongly vested underparts, but ID safest by voice. Call an overslurred, burry drawled *brréeu* or *bzzhiu*; less often a plaintive, slightly upslurred *peéur*, suggesting Eastern Pewee but slightly lower, shorter, less inflected.

imm.

**WILLOW
FLYCATCHER**

**ALDER
FLYCATCHER**

**WHITE-THROATED
FLYCATCHER**

juv.
(Aug–Sep)

**NORTHERN
TROPICAL PEWEE**

juv.
(Jul–Sep)

imm.

EASTERN PEWEE

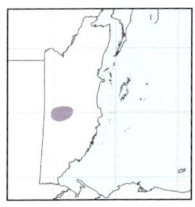

GREATER PEWEE *Contopus pertinax* 17–19cm. Large crested pewee of Mountain Pine Ridge. See genus note. Mainly at mid–upper levels, often in association with mixed flocks in winter. Appreciably larger than other pewees, with tufted crest, bright orange mandible; common call very similar to Olive-sided Flycatcher. Juv. browner above, with buffy-cinnamon wingbars. SOUNDS: Clipped overslurred *bihk*, often in series of 2–4 notes like Olive-sided Flycatcher but more ringing, slightly slower-paced; also repeated persistently, suggesting a *Turdus* thrush clucking. Song year-round a distinctive phrase of 4 slightly plaintive slurred whistles, *whee tee whee'whéu* (or *Jo-sé Ma-ría*); in breeding season often followed by quick rolled *chewdl-it*. STATUS: Uncommon (formerly fairly common) locally in Mountain Pine Ridge. (Mexico and sw. US to nw. Nicaragua.)

OLIVE-SIDED FLYCATCHER *Contopus cooperi* 17–18.5cm. Large, rather compact migrant pewee of forest and edge habitats, especially at higher elevations. See genus note. Perches prominently atop taller trees. Note relatively short tail, peaked hindcrown, white throat with dark-vested underparts; sometimes shows white flank tufts from behind. Two song types may represent cryptic species, not visually separable: **Northern Olive-sided Pewee** *C. [c.] cooperi* (Breeds n. N America); **Western Olive-sided Pewee** *C. [c.]* undescribed? (Breeds w. N America). SOUNDS: Call of both types a clipped, overslurred *bihk*, typically in series of 2–4 notes; migrants sometimes sing: *wot! peeves you* (Western, middle note longest) or *whit free beeh* (Northern, last note longest). STATUS: Uncommon mid-Aug to mid-Nov, late Apr–May; winters locally, mainly at higher elevations, especially Mountain Pine Ridge; wintering birds likely Western. (Breeds N America, winters Mexico to S America.)

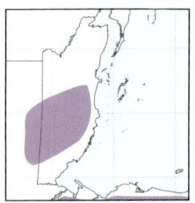

*BLACK PHOEBE *Sayornis nigricans* 15–17cm. Distinctive, medium-size flycatcher usually near water, especially rocky rivers and streams, around bridges, buildings; also towns, villages. Singles or pairs, often on rocks in streams, sally out for insects; often sings in fairly high fluttering flight. Note size, habits, contrasting white belly. Juv. has duller plumage, cinnamon wingbars. SOUNDS: High, sharp, downslurred *siik!* Song high, short piping phrases repeated and alternated *si-ii, s-si-sii, si-ii, s-si-sii…* every 1–2 secs, from perch or in flight. STATUS: Fairly common locally; rare and sporadic wanderer n. of mapped range and to s. coastal belt. (Mexico and w. US to S America.)

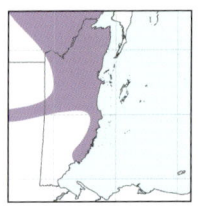

VERMILION FLYCATCHER *Pyrocephalus obscurus* 12.5–14cm. Only Belizean flycatcher in which sexes obviously different. Found in open and semi-open country, especially pine savanna, agricultural areas; often near water. Perches conspicuously on fences, bushes, low trees, sallying and dropping to ground for insects; frequently dips tail like a phoebe. In song flight, male climbs with rapid fluttering wingbeats and plumage puffed out in a ball. Brilliant male unmistakable. Female and imm. distinctive, with dusky breast streaking, whitish eyebrow, variable yellow (1st-year female) to reddish on undertail coverts. Imm. male variably intermediate between adult female and male; attains adult plumage in 2nd year. SOUNDS: High thin *psiep*. Song (can be given on moonlit nights) a few high ticks accelerating into an ascending, short rippled trill, ending with a lower note, *tk tk ti-ti-titisee*, about 1 sec; repeated rapidly in display flight. STATUS: Fairly common to uncommon locally, spreading with deforestation; rarely wanders to larger cayes. (Mexico and sw. US to S America.)

GREATER PEWEE

juv.

OLIVE-SIDED
FLYCATCHER

BLACK PHOEBE

VERMILION FLYCATCHER

male

imm.
female

female

juv.

GENUS *MYIARCHUS* (4 species). Medium-size to rather large flycatchers of wooded and scrubby habitats; nest in tree cavities. Raise bushy crest and puff out throat when agitated or curious (as in response to pygmy owl tooting), slowly jerking head back and forth. Often at fruiting trees, especially in winter. ID often best made by voice; also note prominence of pale wingbars, bill size, tail pattern, face pattern, extent of rusty edging on wings. Juvs. of all species have extensively rusty wings and tail, not easily identified to species; attain adult appearance by early winter.

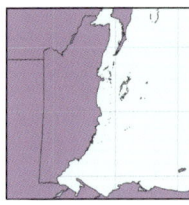

DUSKY-CAPPED FLYCATCHER *Myiarchus tuberculifer* 16–17cm. Small but rather long-billed *Myiarchus*, widespread in varied forested and semi-open habitats, from forest canopy to overgrown weedy fields. Note relatively slender bill, dull pale wingbars, little rusty in tail, voice. Cf. other *Myiarchus*, especially Yucatan Flycatcher, which shares dull wingbars and has somewhat similar calls. **SOUNDS:** Plaintive overslurred *wheéeu*, about 0.5 sec; shorter, more screaming *reeeu* and burrier *wheeer*; shorter note followed by rolled whistle, *whee peeerrrrr*, about 1.5 secs. Dawn song a varied repetition of overslurred whistles and upslurred whip notes every 1.5–3 secs: *wheeéu, wheeéu, wuik! wheébeeu, wuik! wheeéu....* **STATUS:** Fairly common to common. (Mexico and sw. US to S America.)

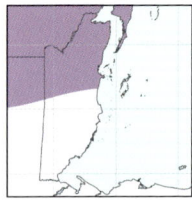

YUCATAN FLYCATCHER *Myiarchus yucatanensis* 17–18.5cm. Medium-size *Myiarchus* of semi-deciduous forest and edge. Mainly at mid–upper levels, often in canopy. Note subtly paler gray lores and face, dull wingbars, contrasting whitish tertial edges, voice. Cf. Dusky-capped Flycatcher. **SOUNDS:** Slurred mournful *wheéeéeu*, about 1 sec; longer, more drawn-out than Dusky-capped Flycatcher, level overall vs. overslurred; drawn-out plaintive whistle run into a bright, slightly emphatic, overslurred ending, *hoooeeEEh*, 1–1.5 secs, much like a song phrase. Dawn song a varied repetition of 2 mournful whistled phrases, every 1.5–2.5 secs: *hooweeéu, hooo'wee-du, hooo'wee-du, hooweeéu, hooo'wee-du....* **STATUS:** Uncommon to fairly common locally. (Mexico to n. Guatemala.)

BROWN-CRESTED FLYCATCHER *Myiarchus tyrannulus* 20.5–22.5cm. Large, relatively big-billed, summer migrant *Myiarchus* of humid forest edge and clearings, semi-open country with hedgerows, pine forest. Note large size, big bill, relatively pale plumage with grayish face, wing and tail patterns, voice. Cf. other *Myiarchus*. **SOUNDS:** Sharply overslurred, clipped *wuík!* and overslurred burry *weéhrr*, latter sometimes in steady series; varied, rhythmic bickering chatters in interactions. Dawn song a measured, varied repetition of quiet pip notes, burry whistles, and short burry chatters, each phrase every 1–2 secs: *whiep, whi'pi-pi-reer, whip, whip, whee-b-beeihr, rreihr....* **STATUS:** Fairly common to common Mar–Aug; uncommon and more local in winter, when usually absent from Ambergris Caye. (Mexico and sw. US to S America.)

GREAT CRESTED FLYCATCHER *Myiarchus crinitus* 19.5–21.5cm. Large, relatively bright migrant *Myiarchus* of humid forest; other habitats in migration, including cayes. Mainly at mid–upper levels, often well hidden in leafy canopy. From other *Myiarchus* by darker gray breast contrasting with brighter yellow belly; also not sharply defined, relatively broad whitish edging to shortest tertial, dull pinkish base to mandible, voice. **SOUNDS:** Upslurred, rich whistled *wheeép*, ending abruptly with slight downward inflection; can be repeated steadily. **STATUS:** Uncommon to fairly common Oct–Mar; commoner and more widespread in migration, Sep–Oct, late Mar–Apr. (Breeds e. N America, winters Mexico to nw. S America.)

DUSKY-CAPPED FLYCATCHER

YUCATAN FLYCATCHER

BROWN-CRESTED FLYCATCHER

GREAT CRESTED FLYCATCHER

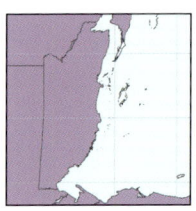

*MIDDLE AMERICAN [BRIGHT-RUMPED] ATTILA *Attila [spadiceus] flammulatus* 19.5–21.5cm. Distinctive flycatcher of humid forest and edge; heard more often than seen. Forages low to high, at times attends ant swarms; often flicks tail upward. Note staring amber eyes, hooked bill, streaked chest, bright tawny rump. Juv. has brownish eyes, broad cinnamon tertial edges. SOUNDS: Sharp nasal *kí-dik* and *kí-di-dik*. Dawn song a far-carrying, often prolonged series of (usually 7–15) loud, paired, slightly screaming whistles, about 2 pairs/sec, typically ending with an overslurred note, sometimes a terminal chip, *whie-dii whie-dii….wheéu chu*, or *knee-deep knee-deep…whoah*, at times faster-paced, with 3-syllable phrases. 'Day song' a slightly overslurred series of (usually 4–10) bright upslurred whistles, about 6/sec, such as *wh-wheéu-wheéu…*, often slowing and followed by *whee-dee-deu*, and variations. STATUS: Fairly common to uncommon. (Mexico to w. Ecuador.)

PIRATIC FLYCATCHER *Legatus leucophaius* 16–17cm. Summer migrant to forest edge, clearings with larger trees; pirates nests of becards, oropendolas. Perches conspicuously in canopy, often on open snags, where may vocalize tirelessly. Distinctive, with stubby bill, dark mask, blurry streaking below; yellow crown patch usually concealed. Juv. has cinnamon wing and tail edgings. SOUNDS: Loud ringing whistle, often followed by 1 or more short twitters, *Sweée, di-di-dit,…*, might bring to mind an overgrown Yellow-throated Euphonia. STATUS: Uncommon to fairly common but local mid-Feb to Aug, stragglers into Sep; exceptional in fall on cayes. (Breeds Mexico to S America, winters S America.)

SULPHUR-BELLIED FLYCATCHER *Myiodynastes luteiventris* 19–21.5cm. Summer migrant to humid forest and edge, pine forest, semi-open areas with tall trees; often at fruiting trees, especially in migration. Mainly at mid–upper levels, often hidden in leafy canopy; nests in tree cavities. Note broad black mustaches meeting under bill, coarse breast streaking, whitish face stripes but yellowish belly (reverse of Northern Streaked Flycatcher), voice; cf. much less numerous Northern Streaked Flycatcher. Yellow crown patch usually concealed. Juv. has pinkish bill base, cinnamon wing and tail edgings. SOUNDS: High, piercing *skweeízik*, suggesting a child's squeezy toy; excited sneezy chatters. Song a fairly quick, rolled *tcheu-wheézilit,…* about every 2 secs. STATUS: Fairly common to common mid-Mar to Sep, stragglers rarely into Oct; more widespread in migration, when occasional on cayes. (Breeds Mexico to Costa Rica, winters S America.)

*NORTHERN STREAKED FLYCATCHER *Myiodynastes maculatus* 20.5–23cm. Summer migrant to humid forest, adjacent clearings with taller trees; often at fruiting trees, especially in migration. Mainly at mid–upper levels; nests in tree cavities. Slightly bulkier and bigger-billed than Sulphur-bellied Flycatcher, with pale bill base, pale chin, finer streaking below extending distinctly to undertail coverts, yellow-tinged face but whiter belly (reverse of Sulphur-bellied), buffy wing edgings; note voice. Yellow crown patch usually concealed. Juv. has cinnamon wing and tail edgings. SOUNDS: Sharp clucking *bihk!* and full-bodied, downslurred *h'chew*; both can be repeated steadily and run into excited chatters. Dawn song a rich, whistled, rippling *wheé-didl-i-eu,…* every 2–3 secs. STATUS: Uncommon and local mid-Mar to Sep. (Breeds Mexico to n. S America, winters Costa Rica to S America.)

MIDDLE AMERICAN ATTILA

PIRATIC FLYCATCHER

watching nest of
Rose-throated Becard

SULPHUR-BELLIED
FLYCATCHER

NORTHERN STREAKED
FLYCATCHER

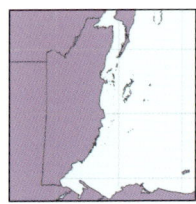

*NORTHERN SOCIAL FLYCATCHER *Myiozetetes [similis] texensis* 17–18cm. Conspicuous, often noisy, medium-size flycatcher of varied edge and semi-open habitats. In pairs or small groups ('social') on wires, at fruiting trees with kiskadees, thrushes. Note small bill, lack of rusty in wings, cf. Great Kiskadee, Boat-billed Flycatcher. Flame crown patch of adult usually concealed. Juv. has cinnamon wing and tail edgings. SOUNDS: Downslurred, slightly piercing *tseéyh*; varied, shrieky and burry bickering series can suggest parakeets, *seéya tortéeya tortéeya…*, etc. Dawn song a varied alternation of downslurred whistles and short burry phrases, *tséu, tséu chirríeu…*, on and on. STATUS: Common to fairly common. (Mexico to Costa Rica.)

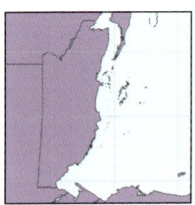

GREAT KISKADEE *Pitangus sulphuratus* 23–25cm. Conspicuous, often noisy large flycatcher of varied edge and semi-open habitats; often near water. Often on roadside wires, at fruiting trees with other large flycatchers, thrushes. Note stout pointed bill, bright rusty wings and tail, voice; cf. Boat-billed and Northern Social Flycatchers. Yellow crown patch usually concealed, lacking on juv. SOUNDS: Loud. Raucous overslurred *reéah* and longer variations, including *kíh káh-réah* ('Kis ka-dee'). Dawn song a varied series of raucous notes and short burry chatters. STATUS: Common to fairly common; occasional in winter on Caye Caulker. (Mexico to S America.)

BOAT-BILLED FLYCATCHER *Megarynchus pitangua* 23–24cm. Large, stout-billed flycatcher of humid forest and edge, adjacent semi-open areas with larger trees. Mainly at mid–upper levels in canopy. Note very stout black bill, lack of bright rusty in wings and tail, voice; cf. Great Kiskadee, much smaller Northern Social Flycatcher. Orange-yellow crown patch of adult usually concealed. Juv. has cinnamon wing and tail edgings. SOUNDS: Grating, harsh, drawn-out *eihrrrrr*, often upslurred at end; screechy bickering chatters, cadence suggesting Northern Social Flycatcher. Dawn song a simple, burry, downslurred *prrriu*, every 1–2 secs. STATUS: Fairly common to common on mainland; absent from cayes. (Mexico to S America.)

KINGBIRDS (GENUS *TYRANNUS*)
(6 species). **Large conspicuous flycatchers of open and semi-open country. Ages/sexes differ slightly, like adult in 1st-year. Juvs. lack concealed yellow or flame crown patches, wing coverts fringed pale cinnamon, fading to whitish. Main sex difference is shape of outer primaries: adult males have narrower, more strongly tapered tips.**

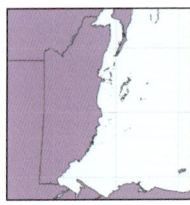

*MIDDLE AMERICAN [TROPICAL] KINGBIRD *Tyrannus [melancholicus] satrapa* 19.5–23.5cm. Conspicuous and familiar large flycatcher of open and semi-open areas, ranchland, towns, parks, lighter woodland, beach scrub, forest clearings and edge; perches on roadside wires, low fences. Cf. Couch's Kingbird (best told by voice). SOUNDS: Varied twittering trills, mainly a short *till-ill-ill-it*, and longer rippling trills that start hesitantly, upslur, then fall at end. Dawn song comprises hesitant tiks run into 2–4-part twittering trills, with pleasing, slightly rippling cadence, *tk tk tk tk widdle-twiddli,…* and variations. STATUS: Common to fairly common, including larger cayes. (Mexico and s. Texas to nw. S America.)

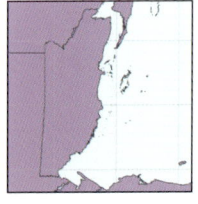

COUCH'S KINGBIRD *Tyrannus couchii* 20.5–24cm. Almost identical to Middle American Kingbird, but favors more forested areas, unlikely in open beach scrub, but the 2 species occur alongside each other, as in towns and villages with larger trees. Often seen on roadside wires, high snags. Couch's averages a shorter bill, brighter plumage, shallower tail fork than Middle American, but best separated by voice. SOUNDS: Clipped *pic*, suggesting Western Kingbird *T. verticalis* but higher and sharper, and burry overslurred drawl, often combined: *pic, pic, vreéehr.…* Dawn song an often hesitant series of bright inflected whistles intensifying and ending with an abrupt downslurred sneeze, *h'week! h'week!…h'wéechu!* and variations; burry drawls thrown in at times. STATUS: Fairly common to common, including larger cayes. (Mexico and s. Texas to Belize.)

NORTHERN SOCIAL FLYCATCHER

juv.

GREAT
KISKADEE

BOAT-BILLED
FLYCATCHER

Kingbird, Kiskadee and
Social Flycatcher

MIDDLE AMERICAN
KINGBIRD

COUCH'S KINGBIRD

EASTERN KINGBIRD *Tyrannus tyrannus* 19.5–21cm. Transient migrant in open and semi-open areas, forest and woodland edge, fruiting canopy. Migrant flocks of 100s can stream overhead all day, especially along the coast in fall. Distinctive (but cf. juv. or molting Fork-tailed Flycatcher), with fairly small bill, blackish head, bold white tail tip; flame crown patch usually concealed. Faded juv. in fall notably duller. SOUNDS: High, shrill, slightly tinny, downslurred *tseih*; high buzzy twitters. STATUS: Common to fairly common mid-Aug to early Nov (stragglers rarely later), fairly common to common mid-Mar to May. (Breeds N America, winters S America.)

GRAY KINGBIRD *Tyrannus dominicensis* 21–23cm. Rare migrant to cayes and open coastal areas, towns; often on roadside wires. Sometimes associates with Middle American and other kingbirds, especially at roosts. Note long heavy bill, relatively long cleft tail, gray upperparts with black mask, whitish underparts; yellow-orange crown patch usually concealed. SOUNDS: Twittering trills similar to Middle American Kingbird but averaging lower, rougher, slower-paced. STATUS: Scarce to uncommon transient mid-Mar to mid-May on cayes, rare on coastal mainland; has bred on cayes and in Belize City, usually departing by Sep; very rare and sporadic in winter. (Breeds Caribbean region, winters se. Caribbean to n. S America.)

SCISSOR-TAILED FLYCATCHER *Tyrannus forficatus* 19–35.5cm. Spectacular kingbird of open habitats, ranchland with hedges and fences, towns. Perches low to high, often on roadside wires; migrants mix readily with other kingbirds and may join roosts of Middle American Kingbird. Distinctive, with pale head and upperparts, variable salmon-pink flush below (darkest on adult male, palest on imm. female); red crown patch usually concealed. In flight, tail tends to splay outward, vs. more lyre-shaped spread of Fork-tailed Flycatcher. Tail longest on adult male. SOUNDS: Clipped, sharply overslurred *pic* and bickering chatters. STATUS: Locally uncommon to rare and sporadic Oct–Apr, occasional from late Sep and into May. (Breeds sw. US and Mexico, winters Mexico to Panama.)

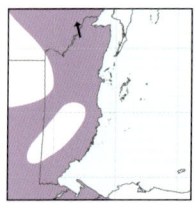

*FORK-TAILED FLYCATCHER *Tyrannus savana* 19–40.5cm. Spectacular kingbird of savanna, marshy areas, ranchland. Perches on ground, low bushes and fences, less often on utility wires. Often in pairs or small groups, locally to 50+ birds; long tail fans in a lyre shape, unlike diverging fork of Scissor-tailed. Distinctive, with black head, long black tail (longest on adult male); yellow crown patch usually concealed. Juv. has sooty gray head. Austral migrants from South America are regular vagrants to North America (mainly Aug–Nov), and may be overlooked in Belize. Differ from resident birds in duskier gray upperparts not contrasting as strongly with wings, slightly lower-pitched call, details of wing-tip structure. Any birds away from normal range should be checked carefully. SOUNDS: High, sharp, slightly metallic *tik* and crackling twitters; song(?) a few *tik* notes accelerating into a short, downslurred gurgle. STATUS: Fairly common to uncommon locally; spreading with deforestation. (Mexico to S America.)

EASTERN KINGBIRD

GRAY KINGBIRD

imm.

SCISSOR-TAILED
FLYCATCHER

juv.

FORK-TAILED FLYCATCHER

MIMIDS (MIMIDAE; 3+ SPECIES) Medium-size, rather long-tailed birds with slender bills, long legs. Mostly terrestrial, but sing from prominent perches. Varied songs often include mimicry, can be loud and protracted. Ages differ slightly, sexes similar; like adult by 1st winter.

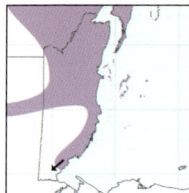

*MAYAN [TROPICAL] MOCKINGBIRD *Mimus [gilvus] gracilis* 23–25cm. Conspicuous in open and semi-open habitats, often perches on roadside wires. Runs well on ground, often with tail cocked. No similar species in Belize: note bold white tail tip, often striking in flight. SOUNDS: Much like Northern Mockingbird *Mimus polyglottos* of North America. Gruff smacking *chk!* often doubled rapidly; low rasping *sshhr*; sharp, slightly rough *tcheu!* Song often prolonged, an unhurried outpouring of varied whistles, clucks, mews, trills, and rasps, typically with 2–6× repetition of phrases and frequent mimicry. STATUS: Common in north, including larger cayes, uncommon to fairly common locally in south; spreading with deforestation. (Mexico to Honduras.)

GRAY CATBIRD *Dumetella carolinensis* 20.5–21.5cm. Skulking migrant mimid of second-growth thickets, forest edge. Forages mainly from ground to mid-levels, but also ranges to canopy of fruiting trees; often dips and fans tail. Distinctive, but in shade can appear all-dark, cf. Black Catbird. SOUNDS: Hoarse mewing *meeah*, can suggest a cat; abrupt rattled *trrrt* in alarm; low clucking *kweh*. STATUS: Common Oct–Apr, a few from late Sep and into May. (Breeds N America, winters Mexico to Panama.)

BLACK CATBIRD *Dumetella (Melanoptila) glabrirostris* 19–20.5cm. Small mimid of second-growth thickets, scrubby forest edge, gardens. Habits much like Gray Catbird, including tail dipping. Note slender bill, uniformly glossy black plumage; reddish eyes show in good light. SOUNDS: Nasal gruff *chehrr*, harsh *rrriah*, and slightly metallic, rasping *tcheeu*; lacks mew call of Gray Catbird. Song a varied, squeaky to scratchy short rambling phrase, can be repeated steadily, often with metallic buzzes thrown in; less often a prolonged rambling warble. STATUS: Common on Ambergris and Caye Caulker, uncommon locally in n. coastal belt; rare and sporadic well inland (year-round, but mainly Nov–Mar) and very rare and sporadic on s. coast. (Mexico to n. Guatemala and Belize.)

THRUSHES (TURDIDAE; 8+ SPECIES) Worldwide family of small to medium-size songbirds with relatively slender bills, often pleasing songs. Ages differ, adult appearance attained within a few months of fledging; sexes similar or different.

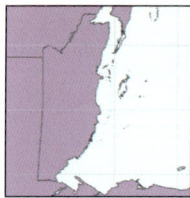

CLAY-COLORED THRUSH *Turdus grayi* 22–25cm. The common large thrush in Belize, widespread in wooded and semi-open habitats. Often feeds on ground, also in fruiting trees. Distinctive but rather plain, with warm brown plumage tones, dull yellowish bill. SOUNDS: Song a rich caroling of mellow whistles, often with slightly lilting or jerky cadence and irregular repetition of phrases. Slurred, wavering mew, *iyeuuh* or *uíreeh*, often rising overall; short soft cluck, often in fairly rapid, slightly laughing short series, *kuh-kuh…*; high thin *siip* mainly in flight. STATUS: Common to fairly common, mainly in lowlands; scarce on Ambergris Caye. (Mexico and s. Texas to n. Colombia.)

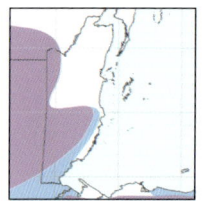

*WHITE-THROATED THRUSH *Turdus assimilis* 22–24cm. Humid forest and edge, adjacent second growth and clearings with fruiting trees and shrubs. Mostly arboreal, but also feeds on ground, usually in shady cover. Distinctive, with slaty-gray to earth-brown upperparts (male averages grayest), darker head with rich yellow bill and eyering (duller on imm.), blackish-streaked throat, white foreneck collar (can be hard to see when birds are overhead). SOUNDS: Song a prolonged, often varied, rich caroling series of mellow whistles, high trills, and fluty notes with frequent 2–3× repetition of phrases, typically fairly leisurely in pace. Calls include a nasal upslurred *rriéh* and *hoo-riéh*; and a low, burry or twangy *urrh*, easily passed off as a frog; high thin *ssi* mainly in flight. STATUS: Fairly common in Maya Mts., mainly above 600m; uncommon in w. Orange Walk; rare and sporadic in adjacent lowlands in winter. (Mexico to Panama.)

MAYAN
MOCKINGBIRD

juv.

GRAY CATBIRD

BLACK CATBIRD

CLAY-COLORED THRUSH

juv.

WHITE-THROATED THRUSH

female/imm.

adult male

EASTERN BLUEBIRD *Sialia sialis* 16.5–18cm. Rare; perhaps no longer occurs in Belize? Attractive thrush known from open pine forest and adjacent habitats in Mountain Pine Ridge. Usually in pairs or small groups; perches prominently on wires, atop bushes, dropping to ground for prey. Flight strong and sweeping; hovers readily in windy conditions. Spotted juv. plumage lost by late summer. SOUNDS: Slightly nasal whistled *cheu* and *chew-it*; low rattled chatter in alarm. Song short rich to slightly burry warbles, often with chatters thrown in. STATUS: Formerly uncommon and local in Mountain Pine Ridge; no confirmed reports in recent years. Decline (disappearance?) probably linked to bark beetle infestation that decimated pines in 2000–2002. (N America to Nicaragua.)

GENUS *CATHARUS* 3+ species of small, spot-breasted migrant thrushes. Ages/sexes similar. Heard more often than seen; migrants may sing in spring. Favor shady forest floor and often elusive, also range into fruiting trees and bushes at mid–upper levels.

SWAINSON'S THRUSH *Catharus ustulatus* 16.5–19cm. Winters in shady humid forest, adjacent habitats; transients may be found in any wooded and forested habitat, including cayes. See genus note. Note buffy face and breast with buff spectacles, plus tone of upperparts (can be difficult to judge in shady understory), cf. Gray-cheeked Thrush, Veery. Comprises 2 groups, perhaps species: **Russet-backed Thrush** *C. [u.] ustulatus* (migrant from w. N America) brighter and warm-toned above; **Olive-backed Thrush** *C. [u.] swainsoni* (migrant from n. and interior w. N America) colder, olive-gray above. SOUNDS: Russet-backed call a mellow rising *whuit*; Olive-backed a higher, sharper, slightly metallic *wuit*. Song often heard in spring migration, a fluty, rich, upward-spiraling warble, 1.5–2 secs. STATUS: **Russet-backed**: scarce mid-Sep to Apr, mainly in south. **Olive-backed**: fairly common to common transient mid-Sep to early Nov, late Mar–May; winter status uncertain. (Breeds N America; Russet-backed winters Mexico to n. Cen America, Olive-backed winters s. Cen America to S America.)

VEERY *Catharus fuscescens* 17–19cm. Transient migrant in varied wooded and forested habitats. Habits much like Swainson's Thrush, and the two species may be found together during migration. From Swainson's Thrush by usually brighter rusty upperparts (birds from w. of breeding range are duller above, much like Russet-backed Swainson's), weak breast spotting, weak pale eyering (shows mainly as a postocular crescent), and smoky-gray flanks. SOUNDS: Downslurred, slightly burry *vheeu*. STATUS: Uncommon to fairly common, mainly mid-Sep to Nov, Apr to mid-May. (Breeds N America, winters S America.)

GRAY-CHEEKED THRUSH *Catharus minimus* 16.5–18.5cm. Infrequently seen transient migrant in varied wooded and forested habitats. Habits much like Swainson's Thrush, and the two species may be found together during migration. From Swainson's Thrush by dull grayish face with poorly defined paler eyering, often colder tones to breast. SOUNDS: Overslurred, slightly nasal *veeu*, similar to Veery. STATUS: Uncommon late Sep–Nov, mid-Apr to mid-May, stragglers into late May; very rare and sporadic in winter. (Breeds n. N America, winters S America.)

WOOD THRUSH *Hylocichla mustelina* 18–20.5cm. Handsome winter migrant of taller humid forest with shady understory; transients may be found in any wooded and forested habitat, including cayes. Habits like *Catharus* thrushes but can be confiding and more conspicuous in areas where acclimated to people. Distinctive: larger and bulkier than *Catharus*, with bright rusty upperparts, bold black spots on white underparts. SOUNDS: Mainly early and late in day, a fairly quick series of (usually 3–6) slightly liquid clucks, *whuit-whuit-whuit…*, and lower *wheh-wheh-wheh….* STATUS: Fairly common to common Oct–Apr; more widespread in migration, mid-Sep to Oct, Apr to mid-May. (Breeds e. N America, winters Mexico to Panama.)

EASTERN BLUEBIRD

juv.

female

male

SWAINSON'S THRUSH

Russet-backed

Olive-backed

VEERY

GRAY-CHEEKED THRUSH

WOOD THRUSH

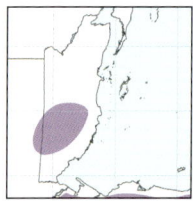

SLATE-COLORED SOLITAIRE *Myadestes unicolor* 19–20.5cm. Slender, long-tailed thrush of cloud forest, ranging seasonally to humid lowland forest. Mainly at mid–upper levels in canopy where perches upright; often still for long periods and overlooked easily if not vocal. No similar species in Belize. Juv. has pale underparts with dark scalloping, like adult by fall; sexes similar. SOUNDS: Ethereal, arresting song often starts hesitant before breaking into a varied series of clear to quavering fluty whistles, sometimes ending with a loose trill, mostly 2–4 secs. Hard nasal *rrank* and buzzier *zzrink*. STATUS: Fairly common at higher elevations in Maya Mts., mainly above 600m; wanders sporadically to lower elevations, mainly in winter. (Mexico to nw. Nicaragua.)

WAXWINGS (BOMBYCILLIDAE; 1 SPECIES) Small Northern Hemisphere family of elegant, rather plump, crested birds. Ages differ, like adult in 1st year; sexes differ slightly.

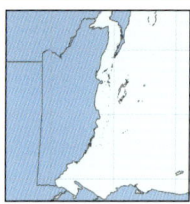

CEDAR WAXWING *Bombycilla cedrorum* 16–17cm. Attractive winter migrant found in a wide variety of semi-open and wooded habitats, from towns to clearings in humid forest. No similar species in Belize. Often in small flocks that fly with slightly undulating motion, wheeling and circling before alighting in fruiting trees; flocks in flight can suggest European Starling *Sturnus vulgaris*. Juv. has blurry dark streaking on underparts; resembles adult by mid-winter. SOUNDS: High, thin, sibilant whistled *sssr*, perched and in flight. STATUS: Uncommon to rare and irregular late Nov–May, mainly at higher elevations; absent some winters. (Breeds N America, winters to Panama.)

WRENS (TROGLODYTIDAE; 9 SPECIES) Mainly New World family known for loud, frequently complex songs but visually elusive habits. All have slender bill, barred wings and tail. Ages similar in most species; attain adult appearance in 1st fall; sexes similar.

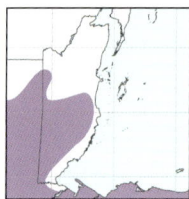

*MESOAMERICAN [BAND-BACKED] WREN *Campylorhynchus [zonatus] zonatus* 18.5–20.5cm. Large, social, noisy wren of humid forest and edge, semi-open areas with taller trees. No similar species in Belize; note boldly banded upperparts and spotted breast, cinnamon flanks and belly; extent of dark barring on underparts variable. Groups of 4–12 birds, less often pairs, forage mainly at mid–upper levels in trees, clambering with agility; joins mixed flocks with orioles, becards, jays. SOUNDS: Gruff rasping chatters, often with rhythmic or slightly jerky cadence; single rasping *shehk*. STATUS: Uncommon to fairly common locally, mainly in Maya Mts., very rare and sporadic wanderer outside mapped range. (Mexico to n. Nicaragua.)

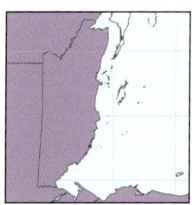

*HOUSE WREN *Troglodytes aedon* 10.5–11.5cm. Small, rather dull wren found in a wide variety of open and semi-open habitats, often around human habitation. Typically skulking and mouse-like, but often sings from prominent perch. Distinctive but rather plain: often has pale eyering but no strong eyebrow or face striping. Juv. has soft dusky scalloping on underparts, soon attains adult appearance. SOUNDS: Song a variably complex, fast-paced, ebullient chortling medley, 2–3 secs; often starts with a few gruff rasps, ends with 2–4 rapid staccato notes or motifs; repeated every few secs. Calls are varied chucks, mews, and rasps; commonly a gruff *chet* and rolled *cherr*, often in hesitant, fairly slow-paced chatters. STATUS: Fairly common locally. (N America to S America.)

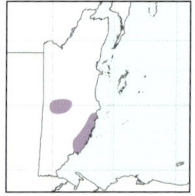

*GRASS [SEDGE] WREN *Cistothorus [platensis] elegans* 11–11.5cm. Small streaky wren of savanna, grassy marshes, damp meadows with taller sedges, rushes. No similar species in Belize; streaking subdued in worn plumage, crown plain on juv. Skulking; moves like a mouse through vegetation, but can pop up, scolding, with tail cocked, and often sings from prominent perch. SOUNDS: Song a varied medley of buzzes, trills, chatters, and chips; song bursts usually 1–2 secs, with pauses of 2–7 secs; same song often repeated several times before changing. Gruff *cheht*, often repeated steadily; low rasping *cherrr*. STATUS: Fairly common to uncommon but local. (Mexico to w. Panama.)

SLATE-COLORED SOLITAIRE

CEDAR WAXWING

MESOAMERICAN WREN

juv.

HOUSE WREN

GRASS WREN

juv.

SPOT-BREASTED WREN *Pheugopedius maculipectus* 12.5–13.5cm. Handsome but rather skulking wren of humid forest edge, second growth, and thickets, especially with leafy and viny tangles. Heard much more than seen; forages low to high in tangles. Distinctive, with striped face and dense black spotting on underparts. Juv. has dusky face and underparts with traces of adult pattern. SOUNDS: Springy, drawn-out rising trill, like a thumbnail on a comb; dry, gruff rapid chatters. Song comprises varied bright series of slurred rich whistles, such as *swee chur-tili-wheechu*, 1.2–1.5 secs; longer in duets; cf. songs of Middle American Wood Wren. STATUS: Fairly common to common. (Mexico to Costa Rica.)

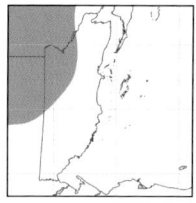

*CAROLINA (WHITE-BROWED) WREN *Thryothorus ludovicianus* 12–13cm. Small wren of semi-deciduous forest understory and edge, adjacent second growth; mainly at low to mid-levels, often in tangles near the ground. Note bold white eyebrow set off by dark eyestripe, black-and-white striping on neck sides, dingy plain underparts, voice; cf. smaller, stub-tailed White-bellied Wren (often in same areas), brighter Plain Wren (limited potential overlap). SOUNDS: Slightly explosive, overslurred buzzy *Bzzzeu*; quieter burry *bzzeihr*; buzzy scolding *zzhi-zzhi...*, and scolding *jeh-jeh....* Song loud, rich, short whistled chants repeated rapidly, typically 4–10×, such as *tch'wee-tch'wee...*, 3–7 phrases/sec; may suggest Northern Cardinal. STATUS: Uncommon and seemingly rather local. (N America to Nicaragua.)

*WHITE-BELLIED WREN *Uropsila leucogastra* 9.5–10cm. Attractive little wren of humid forest understory and edge, adjacent second growth. Ranges low to high, creeping along trunks and branches, often in viny and twiggy tangles; relatively easy to see for a small wren. Conspicuous nests of dried grasses (usually at mid-levels in thorny shrub or small tree) often reveal the species' presence in an area. Distinctive, with blank lores creating 'open' face; also note rather short tail, usually held cocked. Cf. larger, 'stern-faced' Carolina Wren. SOUNDS: Gruff *chuk*, hard rattled *chrrk*, and dry, rather raspy chatters. Song a varied, short, fluty whistled phrase with cheerful, bubbly cadence, such as *ch-widdl-i-eu* or *piddle-ee-oo*. STATUS: Fairly common to common. (Mexico to Honduras.)

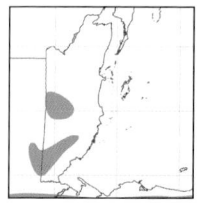

*PLAIN (CABANIS'S) WREN *Cantorchilus modestus* 12.5–13.5cm. Attractive small wren of grassy tangles and second growth in open pine-oak forest, grassy scrub with oaks. Mainly at low to mid-levels, often skulking in dense grassy vegetation. Not especially plain, and in fact subtly attractive, with bold white eyebrow, pale grayish face and breast, bright buffy-cinnamon flanks. SOUNDS: Dry, slightly scratchy *cht*, often in short series, repeated steadily when scolding. Song rich, whistled, short phrases with rippling cadence, usually repeated quickly a few times, often with a lisping introductory note, such as *ss ti-been ti-been, ss ti-been ti-been*; also much higher, thin whistled phrases, such as *tsi-pii-siip* repeated every few secs. STATUS: Uncommon to fairly common locally in Mountain Pine Ridge; scarce and very local in s. foothills. (Mexico to Panama.)

SPOT-BREASTED WREN

juv.

CAROLINA WREN

nest

WHITE-BELLIED WREN

PLAIN WREN

*MIDDLE AMERICAN [WHITE-BREASTED] WOOD WREN *Henicorhina [leucosticta] prostheleuca* 10–11.5cm. Elusive stub-tailed wren of humid forest; heard far more often than seen. Singles or pairs hop on and near forest floor, usually hidden in tangles and foliage. No similar species in Belize, but cf. song of Spot-breasted Wren. Juv. has grayish breast and sides, like adult by fall. **SOUNDS:** Common call a bright, ringing, slightly metallic *peenk* or *biink*, ventriloquial, and could be passed off as a frog. Dry scolding *chek*, low chatters, and a hard, dry, rattling chatter. Songs varied: rich to slightly plaintive whistled short phrases, repeated, *ss chee ree-eu,…* or *hoo-ee hoo'ee-ee,…*, more complex in duets; often introduced by a quiet lisp. Songs typically shorter, more repetitive, less flowing than Spot-breasted Wren. **STATUS:** Fairly common to common; least numerous in northeast. (Mexico to nw. Colombia.)

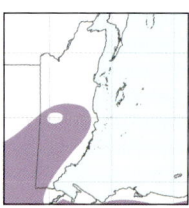

(NORTHERN) NIGHTINGALE WREN *Microcerculus philomela* 10–11.5cm. Small dark wren of humid foothill forest; mainly on or near shady forest floor, where rarely seen unless located by song. Walks with almost constant bobbing motion. Distinctive, but cf. much larger Scaly-throated Leaftosser. **SOUNDS:** Song unmistakable and haunting, a seemingly random, confident to hesitant rising and falling series of plaintive whistles interspersed with thin lisps, such as *hee hoo, hee hoo, hoo hoo hee hoo, ss hoo hee…*; 2–3 notes/sec. Calls include a sharp *chek*. **STATUS:** Uncommon to fairly common but local, mainly in foothills. (Mexico to Costa Rica.)

GNATWRENS, GNATCATCHERS (POLIOPTILIDAE; 3+ SPECIES) New World family of very small, fine-billed, and relatively long-tailed birds. Ages differ slightly (weak juv. plumage soon replaced); gnatcatcher sexes differ in head pattern, at least in breeding season.

*NORTHERN [LONG-BILLED] GNATWREN *Ramphocaenus [melanurus] rufiventris* 12–12.5cm. Distinctive little bird of humid forest edge, adjacent second growth, especially with vine tangles. Forages actively, low to high and often in pairs; tail typically cocked and swung loosely around. Active, can be difficult to see clearly in tangles. **SOUNDS:** Song a dry to vaguely musical, ringing trill, often rising and falling, at times preceded by a quiet tick or chortle, 1.5–3 secs; lower and harsher trills and chatters when agitated; nasal scolding *cheut-cheut*, repeated rapidly. **STATUS:** Fairly common. (Mexico to S America.)

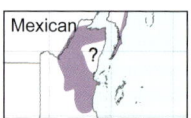

*BLUE-GRAY GNATCATCHER *Polioptila caerulea* 10–11.2cm. Breeds in pine and oak savanna; winter migrants may be found in any wooded and scrubby habitats, but scarce or absent in rainforest. Singles and pairs forage low to high, actively fluttering; tail often cocked and loosely swung about. All plumages have neat white eyering; breeding male has short black eyebrow. 2 groups in Belize may represent species: resident **Mexican Gnatcatcher** *P. [c.] deppei* averages smaller (10–10.5cm), shorter-tailed, and duskier overall, with more contrasting eyering; mandible base gray (year-round?); migrant **Eastern Gnatcatcher** *P. [c.] caerulea* relatively large (10.7–11.2cm) and pale, clean blue-gray above; mandible base pale pinkish in winter. **SOUNDS:** High buzzy mews, averaging higher and thinner in Eastern, lower and harsher in Mexican, often quickly doubled. Song a high, squeaky and buzzy warble, short phrases repeated every few secs or more protracted, with nasal mews thrown in. **STATUS: Mexican:** Fairly common locally; possibly some fall–winter dispersal starting in Aug. **Eastern:** Uncommon to fairly common Sep–early Apr. (N America to Guatemala, winters to Honduras.)

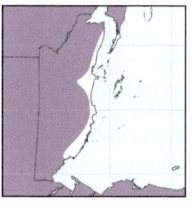

WHITE-BROWED [TROPICAL] GNATCATCHER *Polioptila [plumbea] bilineata* 10–11cm. Humid forest and edge; mainly in canopy, where overlooked easily. Singles or pairs often join mixed flocks of warblers, vireos. Distinctive when seen clearly, with open white face on both sexes, no seasonal change in appearance. Cf. Blue-gray Gnatcatcher. **SOUNDS:** Overslurred, nasal mewing *meéah*, slightly harsher than Blue-gray. Song a fairly high, overall descending, silvery slow trill, 2–3 secs. **STATUS:** Fairly common. (Mexico to w. Peru.)

juv.

MIDDLE AMERICAN
WOOD-WREN

NIGHTINGALE WREN

juv.

NORTHERN
GNATWREN

Eastern

nonbr.
(Aug–Feb)

BLUE-GRAY
GNATCATCHER

female/
nonbr.

Mexican

male breeding
(Feb–Aug)

WHITE-BROWED
GNATCATCHER

female

male

VIREOS (VIREONIDAE; 14+ SPECIES) Mainly New World family of small songbirds. Most vireos resemble heavily built warblers but are, in fact, more closely related to jays and crows. Ages similar or different, soon attain adult appearance; sexes usually similar. Often sluggish, best detected by songs, often repeated tirelessly; calls mainly scolding mews and chatters.

LESSER GREENLET *Pachysylvia decurtata* 10.5–11.5cm. Rather plain little bird of humid forest canopy, occasionally coming lower at edges. Easily overlooked unless voice is known. In pairs or small groups, often with mixed flocks of warblers, other vireos. Note small size, rather short tail, gray head with 'soft' face and whitish eyering, yellowish wash to sides. SOUNDS: Nasal, often persistent scolding *yiih yiih....* Song an unhurried repetition of simple, 2–4-syllable, slightly plaintive whistled phrases, 2nd part lower than 1st, *sípi-chee* or *wíchil-i-wee*, every 2–8 secs. STATUS: Fairly common to common. (Mexico to w. Ecuador.)

TAWNY-CROWNED GREENLET *Tunchiornis ochraceiceps* 11.5–12.5cm. Odd little bird of shady understory in humid forest. Often in pairs with mixed flocks, especially with Stripe-crowned Warblers and Middle American Ant-tanagers in areas with low palms. Note bushy tawny cap, pale eyes in gray face, ochre breast, pinkish legs; cf. female Plain Antvireo (p. 186). SOUNDS: Short series of nasal, overslurred notes, *dwoi-dwoi-dwoi...* and variations. Song a high, plaintive, whining, insect-like whistle, *whiiii*, about 0.5 sec, every 2–3 secs; may be preceded a short liquid trill or chortling chatter. STATUS: Fairly common, especially in south. (Mexico to S America.)

RUFOUS-BROWED PEPPERSHRIKE *Cyclarhis gujanensis* 15–16.5cm. Distinctive, bulky, and striking large vireo of semi-deciduous forest, pine savanna, second growth, brushy hedgerows, mangroves; mainly in drier areas, not in denser evergreen forest. Singles move sluggishly, mainly at mid–upper levels, often with mixed flocks and at fruiting shrubs and trees. No similar species in Belize: note stout bill, gray head with broad rusty eyebrow, dark reddish eyes (browner on imm.). SOUNDS: Song a varied, short warbled series of loud rich whistles, such as *chikee wheer peeripee pee-oo*, or *weer cheery-choo*, every 5–15 secs. Also (female only?) a slightly descending series of (usually 5–12) sad rich whistles, given irregularly, often difficult to trace: *treéu treéu.....* STATUS: Fairly common in north, in smaller numbers s. locally in coastal belt; uncommon on Ambergris Caye. (Mexico to S America.)

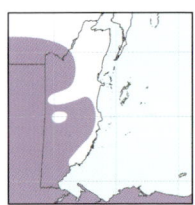

GREEN SHRIKE-VIREO *Vireolanius pulchellus* 13.5–14.5cm. Chunky, bright green bird of rainforest canopy, especially in foothills; heard far more often than seen. Mainly at upper and mid-levels in taller trees, at times with mixed flocks or in fruiting trees with other vireos, tityras, becards; sings persistently from perch in subcanopy. No similar species in Belize. Juv. duller with diffuse yellow face stripes, soon like adult. SOUNDS: Far-carrying song a chant of (typically 3–5) rich whistled notes, *chew chew chew chew*, or *chewy chewy chewy*, every 1–4 secs, at times given tirelessly throughout day; cf. song of Middle American Ant-tanager. Hard, rasping scold, *djehr djehr...* and squeaky rippling chatters. STATUS: Fairly common, mainly in south and foothills. (Mexico to Panama.)

LESSER GREENLET

TAWNY-CROWNED GREENLET

RUFOUS-BROWED PEPPERSHRIKE

juv. **GREEN SHRIKE-VIREO**

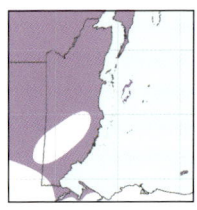

*MAYAN [MANGROVE] VIREO *Vireo [pallens] semiflavus* 11–11.5cm. Small vireo of scrubby woodland, overgrown brushy clearings and thickets, forest edge. Mainly at low to mid-levels, singly or in pairs, at times with mixed flocks. No similar species in Belize: note yellowish lores and small teardrop behind eye, pale amber to pale grayish eyes (dark on imm.), white wingbars, yellowish underparts. Plumage varies from relatively bright yellow-olive above and yellow below to grayish olive above, dingy pale yellowish below. **SOUNDS:** Nasal scolding *sheh-sheh…*, often repeated insistently, much like White-eyed Vireo; drawn-out, slightly overslurred nasal buzz, *jwiehrr*, up to 1 sec. Song notably varied: series of (usually 3–20) twangy or buzzy notes, *jihjih…*, or *j'weih-j'weih…* or *vhizzh-vhizzh…*, 1–2 secs, repeated every few secs; pace varies greatly, often slow enough to count notes but fastest songs approach a ringing nasal trill. **STATUS:** Common to fairly common, especially in north, including Ambergris Caye and Turneffe Is. (Mexico to Nicaragua.)

WHITE-EYED VIREO *Vireo griseus* 11–12cm. Common and attractive winter migrant vireo of brushy woodland, forest edge, second growth, semi-open areas with thickets and scrub. Forages low to high, in winter often with mixed flocks of warblers, other vireos. Distinctive, with gray neck sides, yellow spectacles, staring white eyes (dusky into 1st winter). **SOUNDS:** Nasal scolding *sheh-sheh…*, often repeated insistently. Song can be given in winter, a varied, short burry warble often starting or ending with a sharp chip, such as *tchk! iweedle-iwee chik*, 1–1.5 secs, repeated every few secs. **STATUS:** Common to fairly common late Sep–Apr. (Breeds e. US to Mexico, winters Mexico to Honduras.)

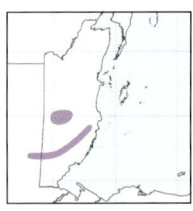

*GRAY-HEADED (CENTRAL AMERICAN) VIREO *Vireo [plumbeus] notius* 12–12.5cm. Local in subtropical pine and mixed forest, especially with oaks. Slightly smaller and duller than Blue-headed Vireo, especially flanks; looks much like allopatric Cassin's Vireo *V. cassini* of Mexico and w. US. **SOUNDS:** Fairly rapid, husky staccato chatter, often slightly descending overall, *ch-ch….* Song varied, rich to slightly burry (usually 2–3 syllable) phrases, repeated with slightly hesitant, question/answer cadence, such as *ch'rieh? ch'reu che-wih, chih-rieh? p'cheu…*, 1 phrase/1–2 secs. **STATUS:** Uncommon to fairly common but local in Mountain Pine Ridge and s. foothills of Maya Mts. (Mexico to Honduras.)

BLUE-HEADED VIREO *Vireo solitarius* 12.5–13.5cm. Scarce winter migrant to varied wooded and forested habitats, especially subtropical pine and pine-oak. Typical birds clearly brighter than Gray-headed Vireo: note blue-gray hood contrasting strongly with silky-white throat, greenish back, yellow flanks; duller birds approach brightest Gray-headed and perhaps not always safely identified. **SOUNDS:** Gruff, scolding, staccato chatters, 1st note often longer and followed by slightly slowing, descending series: *jehh jeh-jeh…*, recalls Yellow-throated Vireo. **STATUS:** Rare late Oct–Apr, most records from Mountain Pine Ridge. (Breeds e. N America, winters to Costa Rica.)

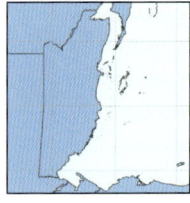

YELLOW-THROATED VIREO *Vireo flavifrons* 13–14cm. Distinctive large migrant vireo of varied wooded and forested habitats, from rainforest to hedgerows with fruiting trees. Note bright yellow spectacles, contrasting white belly, blue-gray rump. Mainly at mid–upper levels, often with mixed flocks and at fruiting trees. **SOUNDS:** Gruff, slow-paced, steady chatter and slightly descending, slowing chatter with longer 1st note, *sheh ch-ch-ch-ch-ch-ch-ch*. **STATUS:** Uncommon to fairly common Sep–Apr; more widespread in migration, late Aug–Oct, late Mar–early May. (Breeds e. N America, winters Mexico to nw. S America.)

MAYAN VIREO

variation

imm.

WHITE-EYED VIREO

GRAY-HEADED VIREO

BLUE-HEADED VIREO

YELLOW-THROATED VIREO

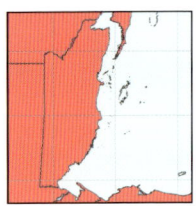

YELLOW-GREEN VIREO *Vireo flavoviridis* 14–15cm. Common summer migrant to varied forested and wooded habitats from gallery forest and gardens to humid forest edge and clearings with taller trees; generally not in dense rainforest. Mainly at upper levels in leafy foliage, coming lower to feed at edges and in fruiting shrubs. Sings tirelessly from canopy; often cocks tail, and raises crown when agitated. Fairly distinctive, with more diffuse face pattern than Red-eyed Vireo, bright yellowish neck sides and flanks, bigger bill. Juv. has dark eyes, duller overall. **SOUNDS:** Gruff downslurred mewing *miehh* and soft dry chatter. Song of varied, burry to slightly nasal chirps given in leisurely, often hesitant manner, *ch-ree, chree, chree, swi ch-ree, chree…*, repeated tirelessly; suggests a House Sparrow that's taken singing lessons. **STATUS:** Common to fairly common late Mar–early Sep, including Ambergris Caye; more widespread in migration, Sep–Oct, Mar–Apr, but rare on cayes. (Breeds Mexico to S America, winters S America.)

RED-EYED VIREO *Vireo olivaceus* 14–15cm. Common transient migrant in varied wooded and forested habitats. Forages low to high, often at fruiting shrubs and trees. Note distinctive face pattern, with neat, thin dark line between broad whitish eyebrow and blue-gray crown; underparts silky whitish overall, with pale yellowish undertail coverts; flanks rarely tinged pale yellow on some fall birds. Cf. Yellow-green and Black-whiskered Vireos. Imm. has brown eyes. **SOUNDS:** Mostly silent in migration; rough, downslurred mewing *rrieh*, averages more drawn-out, rougher than Yellow-green Vireo. **STATUS:** Common mid-Aug to early Nov (stragglers into late Nov), Mar to mid-May, including cayes. (Breeds N America, winters S America.)

BLACK-WHISKERED VIREO *Vireo altiloquus* 14–15cm. Rare transient on cayes, found in scrubby woodland and edge, mangroves. Habits much like Red-eyed Vireo, with which it could occur at fruiting trees and shrubs. Resembles a rather dull, large-billed Red-eyed Vireo without distinct dark line above whitish eyebrow, but with usually distinct dark whisker line. Also cf. Yucatan Vireo. Mostly silent in Belize. **STATUS:** Rare mid-Mar to May on cayes, very rare mid–late Sep. (Breeds Caribbean, winters S America.)

YUCATAN VIREO *Vireo magister* 14.5–15.5cm. Large, big-billed vireo of scrubby forest and edge, mangroves, gardens. Forages with typical sluggish vireo behavior, low to high in fruiting trees and shrubs. Note big bill, thick blackish eyestripe, broad creamy eyebrow, grayish upperparts with variable yellow-green tinge to wing edgings; sometimes shows slight dusky whisker line, cf. Black-whiskered Vireo. **SOUNDS:** Bright nasal *piehk*, sometimes doubled; soft dry chatter similar to Yellow-green Vireo. Song of varied rich phrases given in slightly hesitant manner, often includes doubled notes, *ch'ree, chu-chu, che'u, chu-hu, ch'rieh, chu, chu-hu…*; more leisurely, mellower than Yellow-green Vireo. **STATUS:** Fairly common to common on cayes, uncommon locally on mainland. (Mexico to n. Honduras, Grand Cayman I.)

PHILADELPHIA VIREO *Vireo philadelphicus* 11.5–12.5cm. Fairly small, compact migrant vireo of forested habitats, second growth, hedgerows with fruiting trees. Mainly at mid–upper levels; often joins mixed flocks and visits fruiting trees. Appreciably smaller, smaller-billed, and 'cuter' than Red-eyed and related vireos, with variable yellow wash to throat and breast. Also cf. nonbr. Tennessee Warbler (p. 238). **SOUNDS:** Low gruff *cheh*, often in short series or repeated steadily. **STATUS:** Uncommon to fairly common late Sep to mid-Nov, Mar–May; smaller numbers remain locally through winter. (Breeds N America, winters Mexico to Panama.)

YELLOW-GREEN VIREO

RED-EYED VIREO

BLACK-WHISKERED VIREO

YUCATAN VIREO

PHILADELPHIA VIREO

WOOD-WARBLERS (PARULIDAE; 42+ SPECIES) Large New World family of small insect- and nectar-eating birds, ranging from arboreal to terrestrial, colorful to rather drab. Ages/sexes similar in some species, different in others; juv. plumage typically weak and fluffy, worn briefly (when fed by parents) and rarely seen. Some migrants have marked seasonal plumage changes. ID can be challenging: consider plumage and voice, plus behavior, season, and habitat; migrant species have preferred wintering habitats, but during migration may occur almost anywhere.

Many migrants sing in spring as they move north; some also sing in fall when establishing winter territories. These songs usually heard only briefly in Belize and often rather variable, not described here. Chip call notes are more stereotyped, heard frequently in winter, and often useful for ID. Contact calls ('flight calls') mostly high-pitched, often buzzy, more challenging to distinguish, and mostly not described in the species accounts.

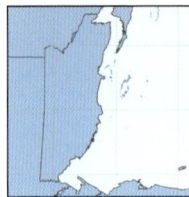

AMERICAN REDSTART *Setophaga ruticilla* 12–13cm. Winter migrant to varied wooded and forested habitats, mangroves, garbage dumps with abundant insects. Forages low to high, often actively fluttering and conspicuous; tail typically fanned to show off big orange (male) to yellow (female/imm.) patches at base. Distinctive, no similar species if seen well. Imm. male like female but with orangey breast patches, often some black spots on head and breast by spring. **SOUNDS:** High sharp *tsip!* similar to Yellow Warbler but slightly higher, sweeter. **STATUS:** Common to fairly common Sep–Apr; more widespread in migration, Aug–Oct, Apr–May. (Breeds N America, winters Mexico to S America.)

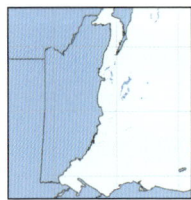

(AMERICAN) YELLOW WARBLER *Setophaga [petechia] aestiva* 11.5–12.5cm. Winter migrant to varied open and semi-open habitats with trees, hedgerows, gardens, mangroves, forest edge and clearings. Forages low to high, mainly at mid–upper levels; territorial in winter. Plumage highly variable, but always rather compact, with pale eyering on open face, pale wing edgings (no wingbars), yellow tail patches (reduced to broad edging on some birds). Cf. Mangrove Warbler. **SOUNDS:** High, sharp, 'generic' warbler *chip!* good to learn for comparison with other species. **STATUS:** Common to fairly common Sep–Apr; more widespread in migration, late Jul–Sep, Apr–May. (Breeds N America to Mexico, winters Mexico to nw. S America.)

MANGROVE [YELLOW] WARBLER *Setophaga petechia* 11.5–12.5cm. Resident in mangroves, ranging occasionally to adjacent habitats. Note shorter primary projection and different voice than migrant Yellow Warbler, which may occur alongside in winter. Adult male Mangrove distinctive, with chestnut head, fine rusty streaking below; imm. and female like Yellow Warbler but often with some rusty markings on face (beware pollen staining on face of migrant Yellow Warblers). **SOUNDS:** Sharp *chuip*, lower and fuller than Yellow Warbler, can suggest Ovenbird. Song a bright, fairly fast, warbled series of sweet chips, 1.5–2 secs; averages richer, lower, and more varied than Yellow Warbler song. **STATUS:** Common to fairly common on cayes and mainland coast. (Mexico to nw. S America.)

PROTHONOTARY WARBLER *Protonotaria citrea* 13–14cm. Distinctive, golden-yellow migrant warbler; winters mainly in mangroves, swampy woodland; in migration also forest and edge, second growth, usually near water. Mainly low in bushes and trees over water, but migrants can be at mid–upper levels in fruiting trees and shrubs. No similar species in Belize: note stout pointed bill, yellow head and breast, plain blue-gray wings, white undertail patches. **SOUNDS:** High tinny *tchín*, less emphatic than Northern Waterthrush. **STATUS:** Fairly common to common late Jul–early Nov, uncommon Mar–Apr; very rare to rare and sporadic in winter. (Breeds e. N America, winters Mexico to nw. S America.)

AMERICAN REDSTART

imm. female

male

adult female
(imm. male similar)

YELLOW WARBLER

drab imm. female
(Sep–Mar)

female

males

MANGROVE WARBLER

imm. female

female

male

PROTHONOTARY WARBLER

imm. female

male

WILSON'S WARBLER *Cardellina pusilla* 11–12cm. Distinctive winter migrant of humid forest and edge, second growth, mangroves. Mainly at low–mid levels in shady understory. Active, sallies frequently for insects, tail often cocked and flipped loosely; joins mixed flocks on occasion. Note beady black eye, habits; often has a black cap; long olive tail lacks white, cf. female Hooded Warbler. Comprises 2 groups: **Western** *C. [p.] pileolata* (from w. N America) brighter, often with orangey lores; both sexes have glossy black cap. **Eastern** *C. [p.] pusilla* (from boreal N America) duller, more greenish overall, female lacks black cap. SOUNDS: Dry *chek*. STATUS: Rare to uncommon Oct–Mar in north, uncommon in south, especially away from coast; more widespread in migration, mid-Sep to Oct, Mar–early May. Relative status of Western and Eastern in Belize awaits study. (Breeds N America, winters Mexico to Panama.)

HOODED WARBLER *Setophaga citrina* 12–13cm. Attractive migrant warbler of humid forest, taller second growth, mangroves. At low to mid-levels, often on forest floor; hops around, frequently flashing white tail spots. Usually independent of mixed flocks, but regular at army ant swarms. Females variable, plainer types told by open yellow face, habits, white tail flashes, voice, cf. Wilson's Warbler. SOUNDS: High tinny *tink*, similar to Eastern Blue Bunting, not as emphatic as waterthrushes. STATUS: Fairly common to common Sep–Apr; more widespread in migration, Aug–Oct, Apr to mid-May. (Breeds e. US, winters Mexico to Panama.)

CANADA WARBLER *Cardellina canadensis* 12–13cm. Transient migrant in varied wooded and forested habitats. At low to mid-levels, often in shady understory where hops and flutters; tail often cocked. Distinctive, with yellow spectacles, dark necklace, blue-gray upperparts; from below, cf. breeding plumage Magnolia Warbler, which has diagnostic tail pattern. SOUNDS: Relatively low, slightly smacking *tchik*. STATUS: Uncommon late Aug–Oct, scarce Apr to mid-May; found mostly on cayes. (Breeds N America, winters S America.)

GENUS *BASILEUTERUS* (2 species). Widespread neotropical genus of forested and scrubby habitats. Sexes similar, often paired year-round; weak juv. plumage held briefly, like adult within a few weeks.

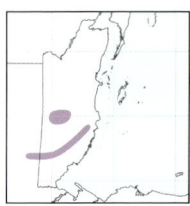

RUFOUS-CAPPED WARBLER *Basileuterus rufifrons* 12–13.5cm. Perky resident warbler of subtropical oak and pine-oak woodland with brushy understory, adjacent overgrown and weedy fields. At low to mid-levels, often in dense brush. No similar species in Belize; note rusty cap, white eyebrow, yellow underparts, cocked tail. SOUNDS: Sharp chipping *chik* often run into excited, chipping series; high sharp *tik*. Song a rapid series of chips, drier at first, then higher, more musical and jangling, with varying pace, about 2–3 secs. Longer songs of 5–8 secs alternate fast chipping and varied jangling, may suggest a siskin. STATUS: Fairly common to common in Mountain Pine Ridge, uncommon and local in s. foothills of Maya Mts. (Mexico to Belize.)

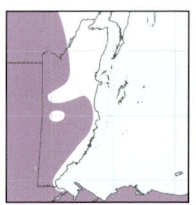

*STRIPE-CROWNED [GOLDEN-CROWNED] WARBLER *Basileuterus culicivorus* 12–13.5cm. Resident warbler of humid forest interior. In pairs or small groups at low to mid-levels in shady understory. Often rather vocal and responsive to pishing; joins mixed flocks, especially with Tawny-crowned Greenlets and Middle American Ant-tanagers in areas with understory palms. Distinctive, with grayish upperparts, striped crown (median crown stripe tawny to yellowish). SOUNDS: Dry *chk* or rattled *trrk*, at times run into chatters suggesting a wren. Song a short sweet warble ending with a strongly upslurred note. STATUS: Fairly common to common, especially in south. (Mexico to Panama.)

WILSON'S WARBLER

female
Eastern

adult
Western

HOODED WARBLER

females

male

CANADA WARBLER

imm. female

male

RUFOUS-CAPPED WARBLER

STRIPE-CROWNED WARBLER

juv.

COMMON YELLOWTHROAT *Geothlypis trichas* 11.5–12.5cm. Widespread winter migrant of marshes, wet fields, mangroves, second growth, scrub; rarely away from water or at least damp vegetation. Often rather skulking, but may respond curiously to pishing. Adult male distinctive, with broad black mask bordered pale bluish gray. Female, especially imm., can be notably drab with little or no yellow on throat and breast, note plain olive head and upperparts, brownish flanks, yellowish undertail coverts. Imm. male has partial black mask in 1st winter, like adult by spring. SOUNDS: Gruff low *tchek* or *chrek*; rattled trill in alarm. STATUS: Fairly common to common Oct–Apr; more widespread in migration, Sep–Oct, Apr–May. (Breeds N America to Mexico, winters s. US to Panama.)

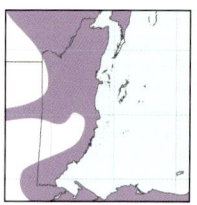

GRAY-CROWNED YELLOWTHROAT *Geothlypis poliocephala* 13.5–14.5cm. Stout-billed warbler of pine and oak savanna, ranchland with rough grass and low bushes; rarely marshes. Mostly skulking in low grassy and shrubby vegetation, but can be curious, perching up on a grass stalk or fence, twitching tail side to side and raising bushy crown; sings from low perch. Distinctive, with stout pinkish bill, dark lores, white eye-arcs, plus habitat and voice. Male averages brighter than female, imm. female dullest. SOUNDS: Bright nasal *chiédl-eu* or simply *chiédl*. Song an unhurried, slightly burry sweet warble, suggesting a *Passerina* bunting, 1.5–3.5 secs; also a descending, slowing series of slightly nasal whistles, 2–3.5 secs, recalls Canyon Wren *Catherpes mexicanus* of N America. STATUS: Fairly common to common locally, especially in coastal belt; occasional on Ambergris Caye. (Mexico to w. Panama.)

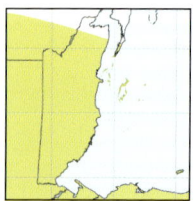

MOURNING WARBLER *Geothlypis philadelphia* 12–13cm. Transient migrant in varied wooded and forested habitats, second growth. Skulking, at low to mid-levels in understory and thickets; hops around like a yellowthroat. Note gray hood, bright yellow underparts, long pink legs. Imm. has narrow, broken pale eyering or narrow pale yellowish spectacles. SOUNDS: Slightly liquid *tchet*. STATUS: Uncommon mid-Sep to mid-Oct, fairly common mid-Apr to early Jun; especially on cayes. (Breeds n. N America, winters Costa Rica to nw. S America.)

Very similar **MacGillivray's Warbler** *Geothlypis tolmei* (12–13cm) has been recorded in Oct and May, likely an overlooked (scarce?) migrant in s. Belize. Note longer tail projection past shorter undertail coverts, voice. All plumages have neatly defined white eye-arcs; imm. has dingy buff throat (not yellowish as on Mourning). Call a slightly wet smacking *tchik*, similar to some variations of Common Yellowthroat but usually higher, sharper.

KENTUCKY WARBLER *Geothlypis formosa* 12.5–13.5cm. Retiring winter migrant of humid forest. Hops around on or near shady forest floor and usually rather shy; best located by call. Distinctive when seen, with yellow spectacles and broad dark sideburns; adult male face blackest, often with blue-gray edging on crown. SOUNDS: Full-bodied, slightly smacking *tchk*, richer and deeper than Ovenbird. STATUS: Fairly common Sep–Apr, especially in south; more widespread in migration, Aug–Oct, mid-Mar to early May. (Breeds e. US, winters Mexico to nw. S America.)

CONNECTICUT WARBLER *Oporornis agilis* 13–14cm. Rare but surely overlooked transient migrant, most likely to be found on cayes in late spring. Notably skulking and difficult to detect; walks (not hops) on or near ground, recalling an Ovenbird; often in shady understory and grassy or weedy cover, but may flush up and freeze in view on low perch. Note stocky build, complete white eyering, walking behavior; also long undertail coverts. Female has dingier gray hood than male. Likely to be silent. STATUS: Probably a scarce spring transient on cayes, mid–late May; one report from Mountain Pine Ridge, 9 May 2013. (Breeds e. US, winters Mexico to nw. S America.)

COMMON YELLOWTHROAT

female

imm. male
(Sep–Mar)

male

imm. female

**GRAY-CROWNED
YELLOWTHROAT**

adult

imms.

MOURNING WARBLER

female

male

male

**MACGILLIVRAY'S
WARBLER**

KENTUCKY WARBLER

imm. female

male

CONNECTICUT WARBLER

male

TENNESSEE WARBLER *Leiothlypis (Oreothlypis) peregrina* 11–12cm. Nonbr. migrant to forested and wooded habitats, second growth; in winter, especially in foothill forest, often in canopy of flowering trees. Face can stain pink from pollen. Note sharp bill, dark eyestripe, whitish undertail coverts. Fall/winter birds bright greenish above, washed yellow on breast, often have narrow pale wingbar, adult male crown grayish, tinged olive; spring adults whiter below, adult male crown blue-gray. **SOUNDS:** Sharp *chik*. **STATUS:** Common to fairly common mid-Sep to Nov, mid-Mar to mid-May, very rare from late Aug and a few into late May; uncommon to scarce and local in winter. (Breeds n. N America, winters Mexico to nw. S America.)

NASHVILLE WARBLER *Leiothlypis (Oreothlypis) ruficapilla* 11–12cm. Rare migrant of wooded and brushy habitats, second growth. Feeds low to high, at times dipping its tail loosely. Face can stain pink from pollen in flowering trees. Note small size, white eyering on gray head, plain olive back (often tinged grayish), mostly yellow underparts. Adult male averages brightest, imm. female dullest. **SOUNDS:** High tinny *tink*. **STATUS:** Rare mid-Sep to Oct, mainly on cayes; very rare Nov to mid-Apr, including mainland. (Breeds N America, winters Mexico to Guatemala.)

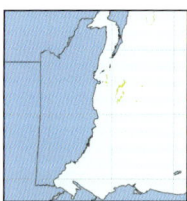

BLUE-WINGED WARBLER *Vermivora cyanoptera (pinus)* 11.5–12cm. Nonbr. migrant to humid forest and edge, second growth. Forages low to high, often in dead-leaf clusters and vine tangles; joins mixed flocks. Distinctive, with dark lores, white wingbars on blue-gray wings, white tail flashes. Hybrids with Golden-winged Warbler rarely reported, show mixed characters. **SOUNDS:** High, slightly buzzy *tssi*, often doubled. **STATUS:** Uncommon to fairly common Sep–Oct, Mar–Apr, with smaller numbers in winter. (Breeds e. N America, winters Mexico to Panama.)

GOLDEN-WINGED WARBLER *Vermivora chrysoptera* 11.5–12cm. Winters mainly in rainforest; migrants may be found in more varied habitats, from beach scrub to semi-open areas with taller trees. Mainly at mid–upper levels, in dead-leaf clusters and vine tangles; joins mixed flocks. Distinctive, with dark mask and throat, yellow wingbars; cf. hybrids with Blue-winged Warbler. **SOUNDS:** High, slightly buzzy *tssi*, often doubled. **STATUS:** Uncommon mid-Sep to Oct, Apr–early May; scarce in winter, mainly in south. (Breeds e. N America, winters Mexico to nw. S America.)

NORTHERN PARULA *Setophaga americana* 10–11cm. Handsome, small migrant warbler of varied wooded habitats, semi-open areas with hedgerows and taller trees, mangroves. Mainly at mid–upper levels, often holds tail cocked. Male distinctive; female/imm. told from Tropical Parula by more extensive gray-blue on head sides (extending through malar region), more distinct white eye-arcs, less extensive yellow on underparts. **SOUNDS:** Sharp *stik*. **STATUS:** Fairly common to common Sep–Apr on cayes and n. mainland, uncommon to rare in south; more widespread in migration, mid-Aug to Oct, Mar–early May. (Breeds e. N America, winters Mexico and Caribbean region.)

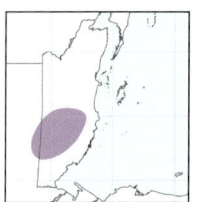

TROPICAL PARULA *Setophaga pitiayumi* 10–11cm. Humid foothill and highland forest. Mainly at mid–upper levels; at times with mixed flocks. Distinctive, with mostly deep blue upperparts, rich yellow underparts, weak white wingbar; lacks white eye-arcs of drabber Northern Parula. Male has more extensive black mask than female, richer orange suffusion to breast. **SOUNDS:** Sharp *stik*, like Northern Parula. Song a variable, high rapid twittering and chipping, at times ending with 1–2 short buzzy trills, 1.5–3 secs. **STATUS:** Fairly common locally at higher elevations in Maya Mts., mainly above 600m; scarce and sporadic at lower elevations. (Mexico to S America.)

TENNESSEE WARBLER

imm.
(Sep–Mar)

face or whole head can
be stained by pollen

male breeding
(Mar–Aug)

imm. female

NASHVILLE WARBLER

male

BLUE-WINGED WARBLER

males

female

male

hybrid Blue-
winged × Golden-
winged Warblers

GOLDEN-WINGED WARBLER

female

male

NORTHERN PARULA

imm. female

adult male

TROPICAL
PARULA

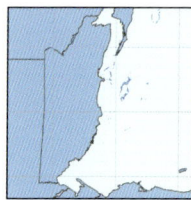

BLACK-AND-WHITE WARBLER *Mniotilta varia* 11.5–12.5cm. Distinctive migrant warbler of varied woodland and forest habitats, especially with larger trees. Creeps along trunks and branches like a nuthatch *Sitta* sp., often with mixed flocks. Behavior distinctive, plus boldly black-and-white striped back, dark centers to undertail coverts. Ad. male has black cheeks, attains black throat Mar–May. SOUNDS: High, slightly liquid *spik*, at times in fairly rapid spluttering series. STATUS: Fairly common to common Sep–Apr; more widespread in migration, late Jul–Oct, Mar to mid-May. (Breeds N America, winters Mexico to nw. S America.)

YELLOW-THROATED WARBLER *Setophaga dominica* 12–13cm. Handsome winter migrant of open and semi-open areas with taller trees (especially coconut palms, pines), forest edge; often creeps around balconies and lights, seeking insects. Usually solitary, not with flocks. Distinctive, with broad black sideburns, white eyebrow and neck patch, long pointed bill. Sexes similar. SOUNDS: High sharp *chik*, recalls Yellow Warbler but averages sharper. STATUS: Fairly common Aug–Apr on cayes and near coast, uncommon to scarce and local inland; more widespread in migration, Jul–Sep, Mar–early May. (Breeds e. N America, winters Mexico to Panama.)

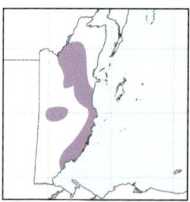

GRACE'S WARBLER *Setophaga graciae* 11.5–12.5cm. Attractive resident warbler of pine savanna, pine forest; rarely away from pine trees. Mainly at mid–upper levels, often foraging in pine-needle clusters; joins mixed flocks. Distinctive, with yellow eyebrow and bib, gray upperparts. Sexes similar. SOUNDS: High sharp *chik*, much like Yellow-throated Warbler. Song a rapid, slightly ringing series of chips, often changes pitch or speeds a little toward end, 1.5–2 secs. STATUS: Fairly common to common. (Mexico to Nicaragua, breeds n. to sw. US.)

MYRTLE [YELLOW-RUMPED] WARBLER *Setophaga coronata* 13–14cm. Fairly large migrant warbler found in open and lightly wooded habitats, such as pastures, pine savanna, coastal scrub. Forages from canopy to ground, usually separate from other warblers; often makes fluttering sallies for insects. Distinctive, but plumage rather variable; note yellow rump, narrow whitish eyebrow and eye-ring, habits. Nonbr. plumages all rather similar, imm. female dullest; breeding female duller than male, with dark grayish not black cheeks. SOUNDS: Fairly strong *tchek*; high *sit*, mainly in flight. STATUS: Sporadic, uncommon to common Nov–Mar in north, generally scarce in south; occasional from mid-Oct and into Apr. (Breeds n. N America, winters US to Cen America.)

BLACK-THROATED GREEN WARBLER *Setophaga virens* 11.5–12.5cm. Attractive winter migrant of wooded and forested habitats, from rainforest to pine woods. Forages low to high; often with mixed flocks. No similar species occurs regularly in Belize: note yellow face with weak olive frame to cheeks, greenish back, yellow tinge to vent. SOUNDS: High sharp *tik*. STATUS: Fairly common to common Oct–Apr; more widespread in migration, Sep–early Nov, Apr to mid-May. (Breeds N America, winters Mexico to nw. S America.)

Hermit Warbler *Setophaga occidentalis* (12–13cm) has been recorded as a winter vagrant in highland pine forest. All plumages have grayish back and lack dark streaking on underparts; adult male striking and distinctive, with golden-yellow head, black bib. Call note like Black-throated Green Warbler.

BLACK-AND-WHITE WARBLER

males

imm. female (Sep–Mar)

nonbr. (Sep–Mar)

breeding (Mar–Aug)

YELLOW-THROATED WARBLER

GRACE'S WARBLER

MYRTLE WARBLER

nonbr. female

male breeding (Mar–Aug)

BLACK-THROATED GREEN WARBLER

female

imm. females (Sep–Mar)

adult male

HERMIT WARBLER

male

MAGNOLIA WARBLER *Setophaga magnolia* 11.5–12.5cm. Attractive winter migrant of wooded and forested habitats, edge, second growth; one of the commonest migrant warblers in Belize. Forages low to high; often with mixed flocks. Fairly active, fluttering in foliage, tail often slightly fanned to show diagnostic pattern on underside: white base with broad black tip. Distinctive: note tail pattern, nasal call, striking difference between nonbr. and breeding plumages. SOUNDS: High, nasal, slightly burry *iehh*, distinct from any other warbler. STATUS: Common to fairly common Oct–Apr; more widespread in migration, Sep–Oct, Apr–May. (Breeds N America, winters Mexico to Panama.)

CAPE MAY WARBLER *Setophaga tigrina* 11.5–12.5cm. Mainly on cayes, favoring forest edge, ornamental gardens, coconut groves, hedgerows with flowering trees. Forages low to high; often around flowering trees and bushes, where territorial. In all plumages note sharply pointed bill, rather compact shape, very fine dark streaking below. Male distinctive (nonbr. plumage like muted version of breeding), with variable rusty cheek patch, white wing panel; female often notably drab, note yellowish neck sides and rump. SOUNDS: Very high, thin, slightly wiry *ti* or *tsi*; lacks a strong chip call. STATUS: Uncommon Sep–early May on cayes; rare to very rare and sporadic on mainland, where possible anywhere but most likely in coastal areas. (Breeds n. N America, winters Caribbean region.)

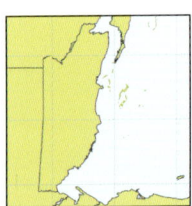

BLACKBURNIAN WARBLER *Setophaga fusca* 11.5–12.5cm. Transient migrant in varied wooded and forested habitats, second growth. Mainly at mid–upper levels, often with mixed flocks. Male stunning and unmistakable, female/imm. have more muted pattern, with pale back stripes, triangular dark cheek patch; yellow throat and breast usually have warm orangey tone. SOUNDS: High sharp, *tsik*, at times with vaguely tinny quality. STATUS: Uncommon late Aug–early Nov (stragglers into Dec); fairly common mid-Mar to May. (Breeds e. N America, winters Costa Rica to S America.)

PRAIRIE WARBLER *Setophaga discolor* 11–12cm. Mainly on cayes, in scrubby woodland, second growth, gardens, overgrown fields. Forages at low to mid-levels; frequently dips tail while hopping around in bushes. All plumages have distinctive face pattern, strongest on ad. male, with broad dark crescent under eye setting off wide yellow eye-arcs; also note weak pale wingbars, tail dipping behavior; imm. has gray cheeks. SOUNDS: High, sharp, slightly smacking *tchik*. STATUS: Uncommon to rare Aug–Apr on cayes, most numerous in migration; rare to very rare and sporadic on mainland, mainly in coastal areas. (Breeds e. N America, winters Caribbean region.)

PALM WARBLER *Setophaga palmarum* 11.5–12.5cm. Conspicuous winter migrant of open beach scrub, gardens, lawns, scrubby woodland. Usually on or near ground, at times in small flocks; hops with near-constant pumping of tail and might better be called 'Pipit Warbler.' Distinctive, with bold pale eyebrow, streaky underparts, yellow undertail coverts, dull yellowish rump. Adult nonbr. often has some rusty in cap; attains bright rusty cap and yellow throat Apr–May. Most birds are relatively dull. **Western Palm Warbler** (race *palmarum*, shown), but at least 2 winter records from Ambergris Caye are of **Yellow Palm Warbler** (race *hypochrysea*), which has underparts suffused bright yellow, even in nonbr. SOUNDS: High sharp *chik* with slight metallic ring. STATUS: Fairly common to common late Sep–Apr (a few from early Sep and into mid-May) on n. cayes, less numerous on n. mainland and s. cayes; rare to very rare and sporadic elsewhere. (Breeds n. N America, winters se. US to Caribbean region.)

MAGNOLIA WARBLER

imm. (Sep–Mar)
(nonbr. similar)

breeding
(Mar–Aug)

male

female

CAPE MAY WARBLER

imm. female
(Sep–Mar)

male breeding
(Mar–Aug)

BLACKBURNIAN WARBLER

female

male

PRAIRIE WARBLER

female

male

PALM WARBLER

Western

nonbr.
(Sep–Apr)

breeding
(Apr–Aug)

CHESTNUT-SIDED WARBLER *Setophaga pennsylvanica* 11.5–12.5cm. Attractive nonbr. migrant of humid forest, second growth. Forages low to high, mainly at mid–upper levels in canopy; joins mixed flocks. Distinctive, with cocked tail and white eyering often recalling a gnatcatcher; note lime-green upperparts and yellow wingbars. Handsome breeding plumage attained Mar–May; female averages duller than male, with less chestnut on sides. SOUNDS: High sharp *chik*, similar to Yellow Warbler. STATUS: Common to fairly common transient, Sep–Oct, Apr–May; uncommon to fairly common in winter, mainly in south and west. (Breeds e. N America, winters Mexico to nw. S America.)

BAY-BREASTED WARBLER *Setophaga castanea* 12–13cm. Transient migrant in varied wooded and forested habitats, second growth. Mainly at mid–upper levels in forest canopy; joins mixed flocks. Rather large and stocky warbler, distinctive in breeding plumage. Imm./fall plumages can resemble much rarer Blackpoll Warbler, but Bay-breasted has a more spectacled expression (vs. pale eyebrow and dark eyestripe of Blackpoll); plainer underparts with little or no dusky streaking, and often tinged buff on flanks; brighter greenish upperparts with thicker white wingbars; and dark feet (yellowish on Blackpoll). SOUNDS: High sharp *chik*, slightly lower and sweeter than Blackpoll Warbler. STATUS: Uncommon to fairly common late Sep–early Nov (stragglers into mid-Dec), Apr–May; most numerous in spring. (Breeds n. N America, winters Panama to nw. S America.)

BLACKPOLL WARBLER *Setophaga striata* 12–13cm. Rare spring transient on cayes, in varied wooded and forested habitats, second growth. Mainly at mid–upper levels in canopy; joins mixed flocks. Rather large, thickset warbler with long wings. Spring male distinctive; on spring female note dark streaking above and below, bright pinkish legs and feet. Fall plumages all similar: olive above, tinged yellow below with diffuse dusky streaking; note pale feet; cf. fall Bay-breasted Warbler. SOUNDS: High sharp *chik*, slightly higher than Yellow Warbler. STATUS: Rare late Apr–May on cayes, very rare mid–late Oct. (Breeds n. N America, winters S America.)

CERULEAN WARBLER *Setophaga cerulea* 11–12cm. Transient migrant in varied wooded and forested habitats, second growth. Mainly at mid–upper levels in canopy; joins mixed flocks. Distinctive, rather compact warbler with fairly short tail often slightly cocked. On imm./female note bold pale eyebrow, unstreaked blue-green upperparts with bold white wingbars. SOUNDS: High sharp *tsik*, slightly higher than American Redstart. STATUS: Uncommon to rare Aug–early Oct; uncommon to locally fairly common late Mar–early May, especially in Maya Mts. Breeds e. N America, winters S America.)

BLACK-THROATED BLUE WARBLER *Setophaga caerulescens* 11.5–12.5cm. Mainly on cayes, in varied wooded and forested habitats, second growth, gardens. Usually at low to mid-levels, often sallying and fluttering in shady lower growth; usually independent of mixed flocks. Male stunning and unmistakable (no seasonal change); female rather drab but distinctive, with narrow whitish eyebrow and lower eye-arc set off by dark cheeks, olive upperparts with small white check at base of primaries (can be absent on some imms.); note call. SOUNDS: Rather low smacking *tchk*, suggests muted Lincoln's Sparrow. STATUS: Uncommon Sep to mid-May on cayes, most numerous in fall; very rare and sporadic on mainland. (Breeds e. N America, winters Caribbean region.)

imm.
(Sep–Mar)

CHESTNUT-SIDED WARBLER

nonbr.
(Sep–Mar)

breeding
(Mar–Aug)

imm.
(Sep–Mar)

BAY-BREASTED WARBLER

female

breeding
(Mar–Aug)

males

nonbr.
(Sep–Mar)

BLACKPOLL WARBLER

breeding
(Mar–Aug)

female

imm.
(Sep–Mar)

male

CERULEAN WARBLER

imm. female

female

male

female

male

BLACK-THROATED BLUE WARBLER

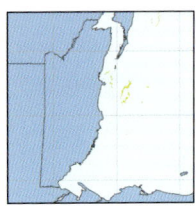

WORM-EATING WARBLER *Helmitheros vermivorus* 12–13cm. Distinctive migrant warbler of humid forest and edge, second growth. Mainly forages from mid-levels to subcanopy, typically in vine tangles and dead-leaf clusters; often joins mixed flocks. Note boldly striped head, rich buffy underparts, dark centers to undertail coverts. Ages/sexes similar. **SOUNDS:** Sharp *tchik*. **STATUS:** Uncommon to fairly common Sep–Mar; more numerous and widespread in migration, mid-Aug to Oct, Mar–Apr. (Breeds e. US, winters Mexico to Panama.)

SWAINSON'S WARBLER *Limnothlypis swainsoni* 13–14cm. Retiring and easily overlooked migrant of humid forest and woodland with leaf litter, tangled understory. Skulking and usually hard to see well, on or near shady forest floor, where hops around, tossing leaves with its bill. Note pale eyebrow, rusty cap, big bill. Ages/sexes similar. **SOUNDS:** Sharp low *chiuk* can suggest Ovenbird but sharper, less often repeated steadily. **STATUS:** Uncommon to rare Sep–Apr in north, scarce and local in south; more widespread in migration, late Aug–Oct, late Mar–Apr. (Breeds se. US, winters Mexico to Belize and Caribbean.)

OVENBIRD *Seiurus aurocapilla* 13.5–14.5cm. Distinctive terrestrial warbler of varied wooded and forested habitats with leaf litter. Walks purposefully on forest floor, tail usually cocked; agitated birds perch low to high, walking along branches and often raising crown feathers. Distinctive, but cf. *Catharus* thrushes (p. 218). **SOUNDS:** Strong deep *tchip* to *chuk*, often repeated steadily when agitated. **STATUS:** Fairly common but low-density Sep–Apr; more widespread and locally numerous in migration, mid-Aug to Oct, Apr–May. (Breeds N America, winters Mexico to Panama.)

NORTHERN WATERTHRUSH *Parkesia noveboracensis* 13.5–14.5cm. Nonbr. migrant of varied wetland habitats, from mangroves to small ponds, lakeshores, slow-moving streams; usually in areas of sluggish or stagnant water. Often rather retiring; walks on ground, pumping its rear end. Little overlap with Louisiana Waterthrush, which favors flowing water; has broader and whiter eyebrow vs. peachy-buff flanks (eyebrow and flanks the same tone on Northern, varying from pale buff to whitish); sparser dark streaking below; bigger bill; deeper bobbing; note voice. **SOUNDS:** Bright metallic *chink*, higher and sharper than Louisiana Waterthrush. **STATUS:** Fairly common to common (especially in mangroves) Sep–Apr; more widespread in migration, mid-Aug to Oct, Apr–May. (Breeds n. N America, winters Mexico to nw. S America.)

LOUISIANA WATERTHRUSH *Parkesia motacilla* 14–15cm. Nonbr. migrant of running clear streams and rivers, especially in foothills, less often lake margins, forest understory; other habitats in migration, but rarely in mangroves. Walks on ground with deep bobbing of rear end, often swung slightly side-to-side; pokes in leaf litter and shallow water. Cf. slightly smaller but relatively longer-tailed Northern Waterthrush, which is typically dingier overall. **SOUNDS:** Full metallic *chiuk*, deeper and less metallic than Northern. **STATUS:** Uncommon to locally fairly common Aug–Mar, especially in hilly country; more widespread in migration, mid-Jul to Sep, Mar–Apr. (Breeds e. N America, winters Mexico to nw. S America.)

**WORM-EATING
WARBLER**

SWAINSON'S WARBLER

OVENBIRD

NORTHERN WATERTHRUSH

LOUISIANA WATERTHRUSH

248

YELLOW-BREASTED CHAT (ICTERIIDAE; 1 SPECIES) Enigmatic songbird long placed within wood-warblers, now treated in its own family. Ages/sexes differ slightly.

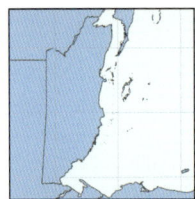

YELLOW-BREASTED CHAT *Icteria virens* 16.5–18cm. Skulking winter migrant of varied scrubby habitats, forest edge, thickets, second growth. Mostly on or near ground in dense vegetation, at times hopping out into adjacent open areas. Distinctive, with bulky build, long tail often cocked, white spectacles, bright yellow bib. SOUNDS: Hard gruff *chrek* and *tek-tek*; low rasping *tcherr*. STATUS: Fairly common to uncommon Oct–Mar; more numerous and widespread in migration, Sep–Oct, Apr–early May. (Breeds N America to Mexico, winters Mexico to Costa Rica.)

NEW WORLD SPARROWS (PASSERELLIDAE; 12+ SPECIES) Large New World family of seed-eating birds with classic 'sparrow' bills; formerly included seedeaters and allies, which are now considered to be tanagers. Ages differ, adult appearance attained within a few weeks; sexes alike.

*OLIVE SPARROW *Arremonops rufivirgatus* 14–16cm. Fairly drab, retiring sparrow of brushy semi-deciduous and pine forest, second growth, overgrown corn fields. Easily overlooked; mostly on ground in shady areas, scratching in leaf litter; sings from low perch. Similar Green-backed Sparrow favors more humid and forested habitats, has brighter greenish upperparts, blue-gray head and breast with black head stripes, pale lemon undertail coverts (vs. buff on Olive Sparrow); note songs. Occasional hybrids may occur. Juv. duller overall with dark streaking above and below, weak head pattern; soon like adult. SOUNDS: High thin *tik*, and sibilant *tsssir*, at times run into excited twittering duets. Song a variably accelerating series of rich chips, typically starts with 1–3 slower-paced chips; mostly 3–5 secs, such as *tcheu, sséu sséu sséu seu-seu-seu....* STATUS: Fairly common. (Mexico to Belize.)

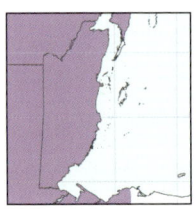

*GREEN-BACKED SPARROW *Arremonops chloronotus* 14.5–15.5cm. Retiring sparrow of humid forest edge and understory, adjacent second growth. Habits much like Olive Sparrow, but mostly segregates by habitat. Cf. Olive Sparrow for differences; beware that occasional hybrids may occur. Juv. duller overall with dark streaking above and below, weak head pattern; soon like adult. SOUNDS: Calls much like Olive Sparrow. Song often richer and sweeter than Olive, not obviously accelerating: a variably paced series of (usually 3–8) rich chips with 1–2 introductory notes, such as *chee, chew-chew-chew-chew-chew*, or *ss, tcheu, chui-chiu-chiu...*; some songs may suggest Green Shrike-Vireo if intro chips are not heard; faster chipping songs can suggest Olive Sparrow, but even-paced. STATUS: Fairly common. (Mexico to nw. Honduras.)

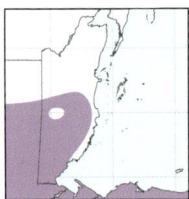

*ORANGE-BILLED SPARROW *Arremon aurantiirostris* 15–16.5cm. Handsome sparrow of humid forest understory, adjacent second growth. Mostly retiring; pairs or singles hop on ground in shady understory; rarely perches in low vegetation except when singing. No similar species in Belize: orange bill often gleams like a beacon in shadowed forest, and white throat can also be striking; imm. bill can be dark into mid-winter. SOUNDS: Slightly rough smacking *tchik!* Rapid, wiry rattling chatter when agitated. Song a very high, squeaky, often slightly jerky warble, 1–4 secs; may suggest Stripe-throated Hermit in quality. STATUS: Fairly common. (Mexico to nw. Panama.)

*MIDDLE AMERICAN [COMMON] BUSH-TANAGER (CHLOROSPINGUS) *Chlorospingus [flavopectus] ophthalmicus* 13.5–14.5cm. Remote cloud forest in south, adjacent shrubby clearings with fruiting bushes. Mainly at low to mid-levels, where active and often noisy in shady understory and edge. Distinctive in habitat and range; note big white postocular spot, yellowish band across breast, habits. Juv. has olive head, lemon tinge and mottling on throat and underparts. SOUNDS: High, sharp, slightly lisping *tsi* and *tsit* notes, sometimes doubled and accelerating into high twitters or trills, 1–2 secs. STATUS: Common at higher elevations in Maya Mts. (Mexico to Honduras.)

YELLOW-BREASTED CHAT

imm.

OLIVE SPARROW

GREEN-BACKED SPARROW

ORANGE-BILLED SPARROW

imm. (Jul–Dec)

juv.

MIDDLE AMERICAN BUSH-TANAGER

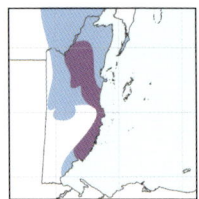

GRASSHOPPER SPARROW *Ammodramus savannarum* 12–12.5cm. Small, stocky, rather short-tailed sparrow, breeding mostly in grassy pine savanna, often with scattered bushes; migrants may occur in any open grassy and weedy habitat. Sings from low perch but at other times notably skulking; occasionally flushes to open perch and freezes to give good views. Note large head, rather short tail, rich buff face and breast with pale eyering. Juv. has streaky breast, ghosting of adult pattern; soon like adult. Migrants average larger, brighter, and paler above than residents. SOUNDS: Calls insect-like, a high soft *tk* or *tik*, and a short, soft dry rattle. Song a high, thin, wiry, insect-like buzz, 1–2 secs, usually 1 or more introductory notes audible at close range. STATUS: Uncommon to fairly common but local resident. Scarce to uncommon nonbr. migrant mid-Oct to Apr, including cayes, whence most records mid-Oct to early Nov. (Breeds N America to Panama, winters n. to s. US.)

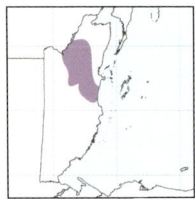

BOTTERI'S SPARROW *Peucaea botterii* 13.5–14.5cm. Rather plain sparrow of pine and pine-oak savanna with taller grassy areas, adjacent grassy fields with scattered bushes. Retiring and elusive when not singing; flushed birds fly low and disappear back into cover; sings from low perch, sometimes in low gliding flight between perches. Plumage rather nondescript, with weak head stripes, plain buffy breast; note long, slightly graduated tail, dark-streaked back. Cf. shorter-tailed Grasshopper Sparrow. Juv. has streaky breast, ghosting of adult pattern; soon like adult. SOUNDS: High, thin, slightly metallic *sik* or *siik*, rather soft. Song a slightly jerky, at times hesitant series of high chips and short trills, about 2 phrases/sec, can be prolonged to 30 secs or more; in full song breaks into an accelerating sweet trill with bouncing ball cadence, such as *ssi, si si-pit sirr si, si-pit see seeu weet weet-weewiwiwiwwiwiwi....* STATUS: Uncommon to fairly common locally. (Mexico and sw. US to nw. Nicaragua.)

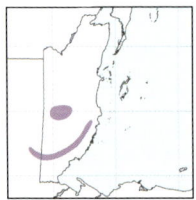

RUSTY SPARROW *Aimophila rufescens* 16.5–20cm. Large bulky sparrow of scrubby second growth, brushy woodland edge, overgrown grassy fields with boulders and scattered bushes, especially in oak and pine-oak zone. Mostly on or near ground, where can be skulking, but sings from low perch; flies heavily if flushed. Appreciably larger and bulkier than other sparrows in Belize; note stout bicolored bill, strong face pattern with black mustache, rusty wings and tail. Juv. streaked overall, with ghosting of adult pattern; soon like adult. SOUNDS: Low gruff *chehr* and gruff scolding chatters, may suggest a wren; run into squeakier chatters in excitement. Song a varied, often rather forceful short phrase of (usually 3–7) bright rich chips, such as *cheeu chik-chik*, or *seeyr seeyr sit-sit-sit-churr*, every few secs. STATUS: Fairly common to common in Mountain Pine Ridge and locally in s. foothills. (Mexico to nw. Costa Rica.)

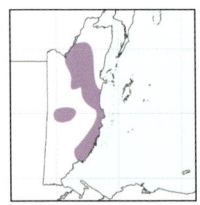

CHIPPING SPARROW *Spizella passerina* 12.5–14cm. Breeds in pine and pine-oak savanna and open pine forest with grassy clearings; migrants could occur in any open and semi-open areas. No similar species regular in Belize, and breeding plumage distinctive (bill becomes black). Imm. and nonbr. variable, can be confused with vagrant Clay-colored Sparrow. Chipping face shows ghosting of breeding pattern, with dark eyestripe extending through lores, pale eyebrow, and variably capped effect; also note gray rump. Clay-colored often buffier overall, with plain lores, dark lower border to cheeks, pale median crown stripe, brownish rump. SOUNDS: High thin *tsip*. Song a rapid dry trill on one pitch, mostly 0.5–2.5 secs; also longer, slower-paced ringing trills, up to 9 secs. STATUS: Fairly common locally. Occasional spring and fall records away from breeding range probably represent migrants from N America. (Breeds N America to Nicaragua, winters n. to s. US.)

Grasshopper Sparrow

juv.

resident

northern migrant

Botteri's Sparrow

juv.

Rusty Sparrow

juv.

Chipping Sparrow

juv.

nonbr. (Sep–Mar)

breeding (Feb–Aug)

SAVANNAH SPARROW *Passerculus sandwichensis* 12.5–14.5cm. Scarce winter migrant to open grassy habitats, from farmland to beaches. Often perches on fence or low bush when flushed, but can be skulking. Note fairly small pinkish bill, relatively short notched tail, fine dark streaking on whitish underparts, and often a yellow tinge to face. **SOUNDS:** High thin *tsit* (often given by flushed birds) and a smacking *tsk*, weaker than smack of Lincoln's Sparrow. **STATUS:** Scarce to uncommon Oct–Apr, most likely to be found in coastal areas and on cayes, but possible anywhere in suitable habitat. (Breeds N America to Guatemala, winters s. US to Honduras.)

LINCOLN'S SPARROW *Melospiza lincolnii* 13.5–14.5cm. Scarce winter migrant to varied weedy, grassy, and brushy habitats, often near water. Mostly skulking, but will respond curiously to pishing, popping up at least briefly and calling. When flushed, usually flies a short distance, low and silently, before slipping back into cover. Note neat dark streaking on rich buff breast, gray face, voice; crown often strongly peaked. **SOUNDS:** Fairly hard smacking *tsk* and high, thin, buzzy *tssir*. **STATUS:** Uncommon to scarce mid-Oct to Apr; regular on cayes during migration. (Breeds N America, winters s. US to Honduras.)

LARK SPARROW *Chondestes grammacus* 15–16.5cm. Vagrant. Distinctive, fairly large, harlequin-patterned sparrow of varied open and semi-open habitats. Likely to be found as singles, often apart from other sparrows. Unlike most sparrows, flight strong and often overhead, calling. Note bold white tail tip, often flashed when landing. Some 1st-years dull, with muted chestnut on face. **SOUNDS:** High, sharp, slightly metallic *sik*, often in flight. **STATUS:** Rare to very rare mid-Sep to Oct, mainly on cayes; exceptional Nov–Feb. (Breeds N America to Mexico, winters n. to sw. US.)

CLAY-COLORED SPARROW *Spizella pallida* 12–13.5cm. Vagrant to open and semi-open grassy and scrubby habitats. Likely to be found as singles, perhaps in association with other sparrows, Indigo Buntings. Slightly smaller and shorter-tailed than Chipping Sparrow, usually with strong buffy tones into mid-winter; best identified by relatively strong face pattern, with pale eyebrow and broad mustache about equally prominent, pale median crown stripe, distinct dark lower border to cheeks, plain lores, brownish rump (vs. gray on Chipping Sparrow). **SOUNDS:** High thin *tsip*, similar to Chipping Sparrow. **STATUS:** Very rare late Sep to mid-Nov, mainly on cayes, where probably annual. (Breeds N America, winters Mexico.)

OLD WORLD SPARROWS (PASSERIDAE; 1 SPECIES) Old World family, one species introduced to New York in 1850 and has spread widely in North and Central America. Ages/sexes differ; resembles adult within 1–2 months, following complete molt.

HOUSE SPARROW *Passer domesticus* 14–15cm. Scarce and local human commensal of towns, urban areas, gas stations. Singles, pairs, or small groups sometimes mix with other birds, but mostly apart. Feeds mainly on ground, often perches on buildings, utility poles, wires. Male head and breast pattern veiled with paler edgings in fresh plumage, bill mostly pale in nonbr. condition. Juv. resembles female, soon attains adult appearance. **SOUNDS:** Varied, mostly rather tuneless chirps and chips, dry chatters, churring calls. **STATUS:** Uncommon and local in south; rare and sporadic n. to Belize City, w. to Belmopan and San Ignacio. (Worldwide; native to Eurasia.)

Savannah Sparrow

Lincoln's Sparrow

Lark Sparrow

Clay-colored Sparrow

House Sparrow

female

breeding male

254

MUNIAS (ESTRILDIDAE; 1 SPECIES) Small seed-eating 'finches' of Old World origin. Populations derived from escaped cagebirds or releases are becoming established locally in New World. Ages differ, sexes similar; attain adult appearance within a few months.

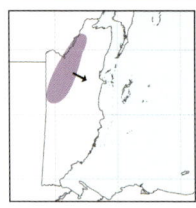

TRICOLORED MUNIA *Lonchura malacca* 10–11cm. Small handsome 'finch' of seeding grassy fields, weedy irrigation ditches and roadsides, rice fields. Often in small groups, locally flocks to 40 or so birds, in same areas as native seedeaters, migrant buntings, and other small weedy field birds, although flocks often keep apart from other species; rather direct 'barreling' flight suggests small compact House Sparrow. Adult distinctive, with bold plumage pattern, small size; juv. rather plain but note large pale bill, rather short and graduated tail. **SOUNDS:** Usually rather quiet; flight call a burry to twangy overslurred nasal *byehh*, singly or in short series. **STATUS:** Uncommon to fairly common locally but spreading; occasional visitor to cayes. First recorded Belize in 2000s. (Native to India.)

FINCHES (FRINGILLIDAE; 8 SPECIES) Almost worldwide family of small, often colorful seed-eating and fruit-eating birds with conical bills. Ages/sexes usually differ, juv. often like female; attain adult appearance within 1st year. Calls and songs notably varied, often including mimicry.

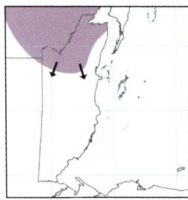

LESSER GOLDFINCH *Spinus psaltria* 9.5–10.5cm. Local resident in open and semi-open areas with scattered trees, towns, villages. Often in pairs or small groups, feeding on or near ground in weedy patches, also in trees and bushes. Note white wing patch and short bill, cf. Black-headed Siskin (normally no range overlap in Belize). Juv. resembles female, with buffy wingbars. **SOUNDS:** High, slightly whiny whistles, often downslurred or overslurred, such as *teeuu*; dry chipping *ch-ch-cht*. Song a varied, often prolonged, fairly rapid jangling warble with frequent mimicry; averages more nasal, buzzier, less rambling than Black-headed Siskin. **STATUS:** Uncommon to fairly common locally but spreading. First recorded Belize in late 1990s. (Mexico and w. US to nw. S America.)

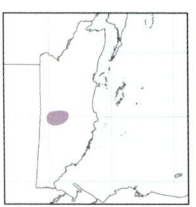

BLACK-HEADED SISKIN *Spinus notatus* 11.5–12cm. Mountain Pine Ridge. Attractive small finch of open pine and pine-oak forest, adjacent clearings and semi-open areas. Feeds mostly in trees, but also low in weedy grasses, brushy fields. All plumages told by big yellow wing patch, relatively slender blue-gray bill, cf. Lesser Goldfinch (normally no range overlap in Belize). Sexes similar, but male averages brighter. 1st-year like juv. but brighter yellowish, often with black mottling on head. **SOUNDS:** Varied nasal and semi-metallic whistles, including a slightly metallic *whéeu* and downslurred *djeeein*; whiny nasal *j'weeih*; and dry *jeh-jeht*. Song a varied, often prolonged, jangling and rambling warble with nasal and metallic notes thrown in. **STATUS:** Fairly common in Mountain Pine Ridge; exceptionally may wander to lowlands. (Mexico to Nicaragua.)

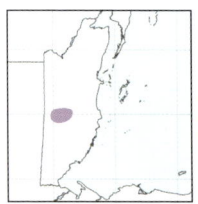

*****RED CROSSBILL** *Loxia curvirostra* 15–16.5cm. Mountain Pine Ridge. Distinctive chunky finch of pine forest; diagnostic crossed bill tips specialized for extracting pine seeds. Flight strong and bounding; most views are of birds overhead, calling. Usually in pairs or small groups; feeds mainly at mid–upper levels in pines; often comes down to drink at creeks and ponds. Male red overall, sometimes mottled with golden; streaked juv. may be seen in any month. **SOUNDS:** Common call, often in flight and usually in series of 2–5 notes, a slightly ringing *drik drik…*; may suggest call of Olive-sided Flycatcher. Song a varied short medley of rich, semi-metallic chirps. **STATUS:** Uncommon nomadic resident in Mountain Pine Ridge; exceptionally may wander to lowlands. (N America to Nicaragua; also Old World.)

TRICOLORED MUNIA

juv.

adult

LESSER GOLDFINCH

female

male

BLACK-HEADED SISKIN

juv.

adult

RED CROSSBILL

female

male

juv.

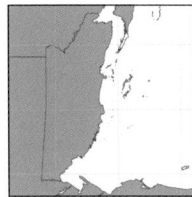

EUPHONIAS (GENUS *EUPHONIA*) (5 species). Small, colorful, tropical finches, formerly considered to be tanagers. Typically in pairs or small groups, often in association with mistletoe. Species readily associate together and also join mixed flocks. Ages/sexes differ, with adult appearance attained in 1st year. Songs and calls notably varied, including mimicry of other species.

YELLOW-THROATED EUPHONIA *Euphonia hirundinacea* 10–11cm. Widespread in semi-open and open areas with fruiting trees, humid forest, and edge. See genus note. Plumage distinctive, but cf. Scrub Euphonia, which readily occurs alongside Yellow-throated in fruiting trees. Juv. resembles female; 1st-year male like female with mostly male head pattern, yellower underparts. SOUNDS: Varied chips, squeaks, short gurgles, and warbles, some burry, often repeated every few secs or mixed into a prolonged song. Common call a slightly nasal, fairly fast-paced *jieh-jieh-jieh*. STATUS: Fairly common to common. (Mexico to w. Panama.)

SCRUB EUPHONIA *Euphonia affinis* 9.5–10cm. Widespread in semi-open and open areas with fruiting trees, hedgerows, forest edge, scrubby woodland. See genus note. Yellow-throated Euphonia slightly larger, male has orange-yellow underparts including throat, female has whitish median underparts with yellowish flanks. Cf. scarce White-vented Euphonia. SOUNDS: Varied short jangling, tinkling, and bubbling phrases, often fairly fast-paced; repeated every few secs or mixed into a prolonged warbling song; quality often slightly plaintive. Common calls include a high, slightly tinny *dee-dee-dee*, upslurred *dwee-dwee-dwee-dwee*, and variations. STATUS: Fairly common to common in north, uncommon and more local in south; occasional on Ambergris Caye. (Mexico to nw. Costa Rica.)

WHITE-VENTED EUPHONIA *Euphonia minuta* 9–10cm. Very small, uncommon euphonia of rainforest canopy and edge, adjacent clearings. See genus note. Small size evident when seen with other species; also note relatively shallow pointed bill. Male has white undertail coverts, richer yellow underparts than Scrub Euphonia; female has pale gray throat contrasting with yellow breast, white median belly and undertail coverts. SOUNDS: Varied chips, twangy warbles, plaintive whistles, and gurgles, repeated every few secs or mixed into a chippering and warbling song. Quality relatively low-pitched and twangy; calls include sharp, slightly smacking, warbler-like *chik!* and a bright, rising *whitzi-chik* phrase often incorporated into songs. STATUS: Scarce to uncommon and local, mainly in foothills; very rare and sporadic n. of mapped range. (Mexico to S America.)

OLIVE-BACKED EUPHONIA *Euphonia gouldi* 10–10.5cm. Distinctive olive-green euphonia of humid forest canopy and edge, adjacent semi-open areas with fruiting trees and bushes. See genus note. Both sexes have diagnostic rusty undertail coverts. Juv. resembles duller version of adult. SOUNDS: Varied burry, nasal, squeaky, and slightly bubbly short phrases, chips, and burry rolled trills, often repeated every few secs or mixed into a prolonged song. Quality relatively rich and burry, including rolled *drrr-rr-rrt*. STATUS: Fairly common to common, especially in south. (Mexico to w. Panama.)

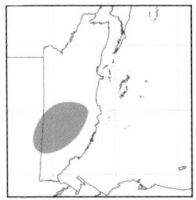

ELEGANT [BLUE-HOODED] EUPHONIA *Euphonia (Chlorophonia) elegantissima* 10–11cm. Beautiful euphonia of humid foothill forest and edge, semi-open areas with trees and shrubs bearing mistletoe. See genus note. Distinctive, with turquoise hood, but duller juv. might be confusing, soon attains blue on crown. SOUNDS: Clipped, descending *teu*, nasal *cheh*, and varied low clucks. Song a pleasant, fast-paced, burbling liquid warble, often prolonged. STATUS: Scarce to uncommon and local in Maya Mts.; rare and sporadic wanderer to adjacent lowlands, and might appear anywhere with mistletoe. Most Belize records are Oct–Apr, but presumed resident. (Mexico to Panama.)

YELLOW-THROATED EUPHONIA

1st-year male

female

male

SCRUB EUPHONIA

female

male

female

WHITE-VENTED EUPHONIA

male

OLIVE-BACKED EUPHONIA

female

male

ELEGANT EUPHONIA

juv.

female

male

NEW WORLD GROSBEAKS AND BUNTINGS (CARDINALIDAE; 20+ SPECIES)

Rather diverse New World family of mostly stout-billed, often colorful, seed-eating and fruit-eating birds. Ages/sexes differ in most species; males of some have distinctive 1st-year plumage. A few n. migrants have seasonal plumage changes. Prior to molecular studies, some species formerly considered as tanagers, others as warblers and New World sparrows.

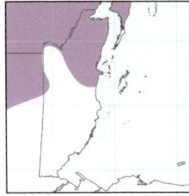

NORTHERN CARDINAL *Cardinalis cardinalis* 20.5–23cm. Distinctive crested grosbeak of scrubby woodland, brushy thickets, overgrown clearings in humid forest. Mainly at low to mid-levels, at times on ground; often rather retiring in thickets but sings from prominent perch. No similar species in Belize. Juv. resembles female but shorter crest lacks red tip, bill dark; 1st-year like adult, bill red by winter. **SOUNDS:** High, sharp, slightly liquid to metallic *tik* or *pik*. Song a series of loud, 1–2-syllable, rich whistled phrases, usually repeated 3–15×; also longer series changing in pattern part way through and sometimes accelerating. Repeats same song several times before switching to another variant. **STATUS:** Uncommon to fairly common, but often rather local. (N America to Belize.)

*NORTHERN BLACK-FACED GROSBEAK *Caryothraustes poliogaster* 18–19cm. Social grosbeak of rainforest and edge, adjacent clearings with taller trees. Typically in small flocks (up to 20 or so birds), roving noisily at mid–upper levels, often at fruiting trees and usually independent of mixed flocks. No similar species in Belize: note contrasting black face, stout bill, blue-gray rump, habits. Juv. duller overall, pattern less sharply defined. **SOUNDS:** Short buzzy rasp followed by (usually 1–3) bright upslurred whistles, such as *t'zzrr whiehk-whiehk*; high upslurred *siup*. Song an unhurried short medley of rich slurred notes, 2–4 secs, often ending with 2–3 strongly downslurred whistles, such as *tsi si-si chu cheu cheu*; also jerkier series, *sih whi-ti wee-chee*. **STATUS:** Fairly common, especially inland and in south. (Mexico to n. Honduras.)

ROSE-BREASTED GROSBEAK *Pheucticus ludovicianus* 18–20cm. Winter migrant to varied wooded and forested habitats, brushy second growth, hedgerows, gardens. Singly or small flocks, the latter especially during migration; feeds on fruits mainly at mid–upper levels, where often sits quietly and is overlooked easily. Adult male striking (plumage heavily veiled brownish in fall–early winter), with big white wing patches; female distinctive, with stout pale bill, coarse dark streaking across breast. Imm. male buffier than female, with bright red underwing coverts. **SOUNDS:** High, squeaky nasal *iihk*, given infrequently. **STATUS:** Uncommon Oct–Apr; more widespread and locally numerous in migration, mid-Sep to mid-Nov, Apr to mid-May. (Breeds N America, winters Mexico to nw. S America.)

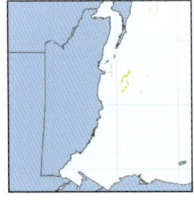

PAINTED BUNTING *Passerina ciris* 12–13cm. Winter migrant to open and semi-open areas with weedy fields, brush, overgrown clearings. Feeds from low in weedy grasses to canopy of fruiting fig trees. Winter flocks rarely more than 20 birds, mixing readily with Indigo Buntings, seedeaters, grassquits. Adult male unmistakable; female and imm. male distinctive, bright plain greenish with paler eyering; imm. female can be rather drab, grayish green. **SOUNDS:** Wet to slightly tinny *plik* or *spik*; infrequently a soft buzz, similar to Indigo Bunting. **STATUS:** Uncommon and often local Oct–Apr; more widespread and numerous in migration, late Sep–Oct, Apr–early May. (Breeds s. US and Mexico, winters Mexico to Panama.)

NORTHERN CARDINAL

female

male

NORTHERN BLACK-FACED GROSBEAK

ROSE-BREASTED GROSBEAK

imm. male
(Sep–Mar)

female

male
(Sep–Mar)

male
(Mar–Aug)

PAINTED BUNTING

1st-year female

adult female
(1st-year male similar)

male

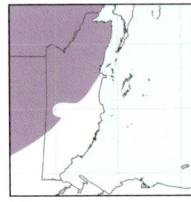

*EASTERN BLUE BUNTING *Cyanocompsa parellina* 13–14cm. Fairly chunky, stout-billed bunting of semi-deciduous forest, adjacent second growth. Usually singles or pairs low in brush or shady understory, not in flocks or in open situations. Male can look blackish, but blue highlights show in good light; imm. male can be vaguely mottled, cf. smaller-billed Indigo Bunting. Female and juv. rich brown overall, cf. stouter-billed female Thick-billed Seedfinch, much larger Blue-black Grosbeak. Also cf. scarce Blue Seedeater, which is rarely away from bamboo. **SOUNDS:** Metallic *tink* or *chink*, suggests Hooded Warbler. Song a short, sweet, slightly rambling warble, intro note often overslurred and subtly separate from main song, 1–2 secs. **STATUS:** Fairly common. (Mexico to nw. Nicaragua.)

*BLUE SEEDEATER *Amaurospiza concolor* 11.5–12.5cm. Scarce denizen of forest edge and second growth with stands of bamboo, perhaps especially along rivers. Singles or pairs forage almost exclusively on seeding bamboo, mainly at low to mid-levels; sings from cover while foraging, also from prominent perch. Subtly distinctive but easily overlooked; note grayish bill, uniform plumage; cf. Eastern Blue Bunting, stouter-billed Thick-billed Seedfinch (p. 270). Imm. resembles female but 1st-summer male often has variable slaty-blue patches on head and body. **SOUNDS:** High, sharp, slightly metallic *tsik* or *sik*; also thin twitters. Song a high, fairly rapid, pleasant warble, 1–1.5 secs, higher and thinner than Eastern Blue Bunting. **STATUS:** Scarce and local; likely more widespread than shown. (Mexico to S America.)

BLUE-BLACK GROSBEAK *Cyanoloxia (Cyanocompsa) cyanoides* 17–18.5cm. Dark, very stout-billed grosbeak of humid forest and edge, adjacent second-growth thickets. Often in pairs, mainly at low to mid-levels in leafy foliage and shady understory; usually apart from mixed flocks. Relatively large size and massive bill distinctive, along with habitat and habits; cf. Eastern Blue Bunting. Juv. like female but darker; 1st-year male often has scattered blue-black feathers in 1st summer. **SOUNDS:** Sharp, fairly loud, slightly nasal to squeaky *plik* and *pli-dik!* Song fairly loud, a slightly descending slow warble of rich slurred whistles, 2–3 secs; sometimes fades into a soft squeaky ending. **STATUS:** Fairly common to common, especially in south. (Mexico to nw. Peru.)

INDIGO BUNTING *Passerina cyanea* 12–13cm. Winter migrant to open and semi-open areas with weedy fields, overgrown clearings, grassy lawns; feeds from low in weedy grasses to canopy of fruiting trees. Flocks locally can number 100s, mixing readily with seedeaters, grassquits. Breeding male stunning and distinctive, but blue reduced to variable mottling in winter; imm. male attains blue plumage over winter but never solidly blue. Female and 1st-winter warm brown overall with variable dusky breast streaking; some fresh adult females rich brown with subdued streaking, and imm. wingbars can fade to whitish. Cf. female Blue-black Grassquit (p. 270), which is smaller with graduated tail, often a contrasting whitish eyering. **SOUNDS:** Wet *plik* and buzzy *zzzrt*. **STATUS:** Fairly common to common Oct–Apr; more widespread and locally numerous mid-Sep to mid-Nov, mid-Mar to mid-May. (Breeds s. N America, winters Mexico to Panama.)

BLUE GROSBEAK *Passerina caerulea* 16.5–17.5cm. Winter migrant to open and semi-open areas such as weedy fields, marshes, overgrown forest clearings, corn fields. Singles or small flocks associate readily with other *Passerina* buntings, seedeaters. Note very stout bill, broad cinnamon wingbars; hindcrown often strongly peaked. Cf. appreciably smaller Indigo Bunting. Juv. resembles female with brighter pale cinnamon wingbars, which can fade to whitish by late fall. 1st-summer male plumage variably intermediate between female and adult male. Adult male in fresh plumage has blue veiled with cinnamon-brown edgings, wearing off to reveal waxy royal blue by spring. **SOUNDS:** Strong metallic *chink*; wet buzzy *zzzir*, lower-pitched than Indigo Bunting buzz. **STATUS:** Uncommon to fairly common Oct–Apr; more widespread and locally numerous Sep–Oct, mid-Mar to mid-May. (Breeds US to nw. Costa Rica, winters Mexico to Panama.)

EASTERN BLUE BUNTING

female

male

BLUE SEEDEATER

female

male

BLUE-BLACK GROSBEAK

female

male

INDIGO BUNTING

female

male (Sep–Apr)

male (Apr–Sep)

BLUE GROSBEAK

female (1st-year male similar, Aug–Feb)

adult male (Sep–Feb)

adult male (Feb–Aug)

DICKCISSEL *Spiza americana* 14.5–16cm. A rather sparrow-like migrant cardinalid found in crop and weedy fields, damp grassy areas with scrubby bushes. Singles associate with other cardinalids, especially Blue Grosbeak, also with seedeaters. Migrating flocks can number 100s, flying overhead in undulating swarms that suggest blackbirds. Breeding male distinctive, with yellow face and breast, black bib; female and imm. have ghosting of male pattern, stout bill, streaked flanks. Cf. female House Sparrow (p. 252). **SOUNDS:** Wet buzzy *zzzrt*, mainly in flight, suggesting emphatic rough-winged swallow; full, slightly liquid *fwit*. Spring males sometimes sing: jangling and chipping medleys ending with a bright buzzy *dik-cizz-l*. **STATUS:** Sporadically fairly common to common, mid-Aug to mid-Nov, Mar–May; commonest in spring and near coast, with peak numbers mid-Sep to Oct, mid-Apr to early May; uncommon and sporadic in winter in s. Toledo. (Breeds N America, winters Mexico to n. S America.)

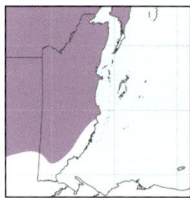

GRAY-THROATED CHAT *Granatellus sallaei* 12.5–13.5cm. Attractive small bird of semi-deciduous forest, brushy woodland, damp thickets in patches of drier forest. Singles and pairs forage at low to mid-levels in tangled undergrowth; usually independent of mixed flocks but often attends army ant swarms. Hops with tail cocked, fanned, loosely waved about. Distinctive, but female can be puzzling; note habits, beady dark eye in plain buffy face, long expressive tail. **SOUNDS:** Soft *whet* or *whiut*, may suggest a *Passerina* bunting. Song a sweet, slightly sad warble, overall descending, 1–1.5 secs. **STATUS:** Uncommon to fairly common in north, uncommon to scarce and local in s. foothills. (Mexico to Belize.)

GENUS *HABIA* (2 species). Rather plain cardinalids (formerly considered as tanagers), typically found in small groups foraging at lower and mid-levels of shady forest understory. Often rather vocal.

*MIDDLE AMERICAN [RED-CROWNED] ANT-TANAGER *Habia [rubica] rubicoides* 17–20cm. Humid forest, especially in hilly terrain. In pairs or small groups at low to mid-levels in leafy understory, often with Tawny-crowned Greenlet and Stripe-crowned Warbler in areas with understory palms. Can occur alongside Red-throated Ant-tanager at army ant swarms: Red-throated darker overall with slightly longer bill, low rasping calls; male has dark lores, female has buff throat. Erectile crown stripe of Middle American not always visible. Juv. (plumage held briefly) overall sooty brown; 1st-year male resembles female. **SOUNDS:** Often utters squeaky and spluttering conversational chips and chatters. Song a simple arrangement of 2–6 rich to plaintive, often slightly burry whistled notes, such as *chieh chieh chéu chéu chéu*, repeated every few secs; shorter songs often repeated 2–5 times in series, such as *chieh-choo chieh-choo…*; typically lacks jerky rhythm of Red-throated Ant-tanager. **STATUS:** Fairly common to common; most numerous at higher elevations. (Mexico to Panama.)

*RED-THROATED ANT-TANAGER *Habia fuscicauda* 18.5–21.5cm. Humid forest and edge, adjacent second-growth thickets. Habits much like Middle American Ant-tanager but often forages lower, and less of a forest-based bird. See under Middle American for ID. Juv. overall sooty brown; 1st-year male resembles female. **SOUNDS:** Often utters low rasping *shehh-shehh…*, also hard, dry, rattled *ch-ch-cht*. Song varied, typically a rhythmic (usually 2–6×) repetition of simple 2–5-syllable, rich, loud whistled phrases with distinctive jerky rhythm, such as *chu-ree'choo chu-ree'choo…*; averages more complex than Middle American Ant-tanager, although sometimes simply *ch-choo ch-choo…*, but still with distinctive jerky rhythm. **STATUS:** Common to fairly common; least numerous at higher elevations. (Mexico to Nicaragua.)

DICKCISSEL

imm. female

male breeding

GRAY-THROATED CHAT

female

male

MIDDLE AMERICAN ANT-TANAGER

female

male

female

male

RED-THROATED ANT-TANAGER

GENUS *PIRANGA* (6 species). Arboreal, boldly colored and patterned cardinalids favoring forest canopy, where easily overlooked if not vocal. For ID note overall plumage patterns and colors, bill size and color, voice. Formerly treated as tanagers and sometimes considered to include genus *Spermagra*, which differs in structure, habits, voice, juvenile plumage pattern, molts.

SUMMER TANAGER *Piranga rubra* 17–19.5cm. Winter migrant to varied wooded and forest edge habitats, from mangroves to pine forest. Typically singles, at mid–upper levels; often at fruiting trees, and sallies for wasps from prominent perches. Note long stout bill, plumage tones; often holds tail cocked, crown feathers raised. Hepatic Tanager has gray bill with distinct notch on cutting edge, grayish cheeks and back, calls distinct. Smaller-billed female Scarlet Tanager more greenish. Female Summer typically plain ochre-yellow, but some blotched rusty red. 1st-year male like female, often with red patches in 1st summer. SOUNDS: Rolled, slightly wet, soft chattering *pí-tuh-ruk*, or *pí-tuh-t-ruk*, typically 2–4-syllables, rarely run into a spluttering rattle. STATUS: Fairly common Oct–Apr; more widespread and locally numerous in migration, Sep–Oct, Apr to mid-May. (Breeds N America and Mexico, winters Mexico to S America.)

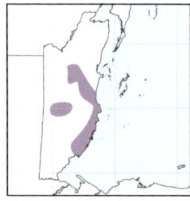

HEPATIC (NORTHERN HEPATIC) TANAGER *Piranga [flava] hepatica* 18–19.5cm. Pine savanna and pine-oak forest, adjacent clearings. Mainly singles and pairs at mid–upper levels, but also feeds readily on ground, tail often cocked and loosely flipped side to side. Fairly distinctive but cf. Summer Tanager, which can occur alongside in winter; note gray bill and variable silvery gray cheeks of Hepatic. Juv. like paler female with extensive dark streaking, pale buffy wingbars; soon attains greenish plumage, 1st-year male like female. SOUNDS: Low clucking *chep*; nasal upslurred *weink*, often in flight. Remarkably, populations in northern Central America are not known to sing. STATUS: Fairly common to common. (Mexico and sw. US to Nicaragua.)

SCARLET TANAGER *Piranga olivacea* 16.5–18cm. Transient migrant in varied wooded and forested habitats, hedgerows, coastal scrub. Singles and small groups forage mainly in canopy, often at fruiting trees. Male distinctive, with contrasting blackish wings, and stunningly scarlet in spring; female smaller and smaller-billed than Summer Tanager, with greenish-yellow vs. mustard-yellow plumage tones, blank lores. Some show narrow pale wingbars, cf. Western Tanager. SOUNDS: Often silent. Low clipped *chk*, can be followed by burry whistled *vrirr*. STATUS: Uncommon to sporadically fairly common mid-Sep to Nov, mid-Mar to mid-May. (Breeds e. N America, winters S America.)

SUMMER TANAGER

females

imm. male
(Dec–Aug)

male

HEPATIC TANAGER

female

male

SCARLET TANAGER

female

male
(Sep–Feb)

male
(Mar–Aug)

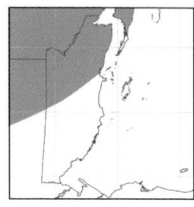

ROSE-THROATED TANAGER *Piranga roseogularis* 15–16.5cm. Distinctive 'tanager' of semi-deciduous forest and edge. Singles and pairs feed mainly at mid–upper levels, often in canopy; regular with mixed flocks. Note stout bill, strong but broken whitish eyering; male handsome but female rather under-stated, with yellow blush in place of rose-red. 1st-year male like female, but often some red on wings in 1st summer. **SOUNDS:** Gruff, downslurred rasping *rreh* or *rrehr*, and nasal, downslurred mewing *nyieh*, may suggest a catbird. Song a short, rich warble of slightly burry phrases; suggests Summer Tanager but faster, more slurred, usually shorter. **STATUS:** Uncommon to fairly common locally. (Mexico to Belize.)

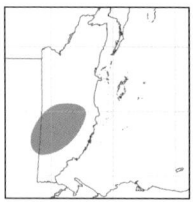

WHITE-WINGED TANAGER *Spermagra (Piranga) leucoptera* 13.5–14.5cm. Attractive small 'tanager' of humid foothill forest and edge, adjacent clearings and second growth with taller trees. Pairs and small groups glean in foliage at mid–upper levels, often with mixed flocks. Smaller than *Piranga*, with smaller bill, clean white wingbars, distinct voice. Cf. appreciably larger Flame-colored Tanager. Juv. resembles female, male soon attains red to orange-red head and body plumage. **SOUNDS:** Short series of (usually 1–5) high, slightly tinny upslurred whistles preceded by quiet chip, *t'sieh sieh*, and variations. High, slightly liquid chipping *tik*, may be repeated steadily; high spluttering twitter in flight. Song a high, slightly squeaky short warble, 1.5–3 secs, such as *si si-see-see chu*. **STATUS:** Fairly common to uncommon in Maya Mts., mainly above 500m; wanders to lower elevations mainly in fall–winter, when occasional n. of mapped range. (Mexico to S America.)

FLAME-COLORED TANAGER *Piranga bidentata* 18–19cm. Very local in remote, humid highland forest. Singles and pairs forage quietly in canopy, at times with mixed flocks. Distinctive, with dusky cheeks, whitish wingbars, dark streaking on back; also note relatively big grayish bill, cf. Western Tanager. Juv. (plumage held briefly) paler than female, with dark streaking below; 1st-year male like bright female, some with variable reddish wash and patches on head and breast. **SOUNDS:** Hard rolled *ch-t-ruk* and *p-terruk*, similar to Western Tanager but slightly drier, sharper. Song comprises 3–5 burry phrases with unhurried, slightly jerky cadence, *chik churree chuwee*, and variations, much like Western Tanager. **STATUS:** Uncommon at higher elevations on Mt. Margaret and nearby peaks; first discovered 1994 in Belize, and may occur elsewhere in Maya Mts. (Mexico to Panama.

WESTERN TANAGER *Piranga ludoviciana* 16.5–18cm. Rare winter migrant to varied wooded and forested habitats. Mainly at mid–upper levels, often in canopy of fruiting trees. Note pale bill, bold pale wingbars, unstreaked back. Some imms. very drab overall, others bright yellow below. **SOUNDS:** Rolled *ch-t-ruk*, slightly lower and faster paced than often more spluttering call of Summer Tanager; nasal upslurred *weink*, often in flight, similar to Hepatic Tanager. **STATUS:** Rare to very rare and irregular Oct–Apr; records scattered throughout the country. (Breeds w. N America, winters Mexico to Costa Rica.)

ROSE-THROATED TANAGER

female

male

WHITE-WINGED TANAGER

female

male

imm. male
(Dec–Aug)

FLAME-COLORED TANAGER

female

male

WESTERN TANAGER

female

male
(Sep–Feb)

male
(Mar–Aug)

TANAGERS (THRAUPIDAE; 22 SPECIES) Very large, diverse Neotropical family. Traditionally, a tanager was a small, colorful, fruit-eating bird. Molecular studies have found, however, that many species formerly treated in other families are actually tanagers that have diversified to fill varied niches, much like the iconic Galapagos 'finches'—which are also now tanagers! Meanwhile, numerous former 'tanagers' have been moved into other families. Tanagers now include saltators, grassquits, seedeaters, and honeycreepers. Ages/sexes differ or similar, attain adult appearance within 1st year, often within a few weeks.

GENUS *SALTATOR* (3 species). Large, stout-billed tanagers, formerly considered as cardinalids. Ages differ, sexes mostly similar. Often unobtrusive, quietly eating leaves, flowers, fruit; best detected by voice.

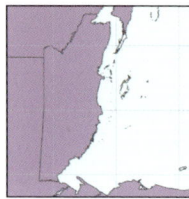

CINNAMON-BELLIED [GRAYISH] SALTATOR *Saltator [coerulescens] grandis* 21–23.5cm. Distinctive tanager of varied wooded, edge, and second-growth habitats. Feeds low to high, from leafy vine tangles to fruiting canopy. Distinctive, with stout black bill, white eyebrow; imm. has strong olive suffusion, pale yellowish eyebrow. SOUNDS: Varied arrangements of rich whistles, chips, and warbles, at times prolonged; often including a drawn-out, strongly upslurred whistle *teu-whieeeeeh* and a quavering trill *chur-r-r-r-r*. Common call a high, slightly squeaky or tinny *ssii* or tinny *tsii*. STATUS: Fairly common to common; occasional on Ambergris Caye. (Mexico to w. Panama.)

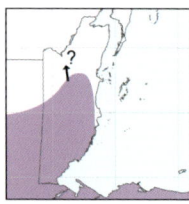

BUFF-THROATED SALTATOR *Saltator maximus* 20.5–23cm. Smallest saltator in Belize, found in humid forest and edge, second growth. Forages low to high; usually in pairs and often rather unobtrusive, feeding quietly on flowers and in fruiting trees. Note gray head, buff throat patch; cf. appreciably larger and often noisier Black-headed Saltator. SOUNDS: Rather quiet. High, slightly tinny or sibilant *tsii*. Song a pleasant, slightly tinny, overall downslurred slow warble, about 1 sec. STATUS: Fairly common to common in south, scarce and local at n. edges of range; may be expanding north. (Mexico to S America.)

BLACK-HEADED SALTATOR *Saltator atriceps* 25.5–28cm. Large, often noisy tanager of humid forest canopy and edge, adjacent clearings with trees and shrubs. Typically in small groups at mid–upper levels; behavior ranges from quiet and shy to loud and obnoxious; joins mixed flocks with orioles, Green Jays. Distinctive, with big white throat patch, large size, loud calls; cf. appreciably smaller and quieter Buff-throated Saltator. SOUNDS: Often rather noisy. Sharp smacking barks and chuckling chatters, often run together. STATUS: Fairly common to common. (Mexico to w. Panama.)

GRASSLAND YELLOW-FINCH *Sicalis luteola* 10–11.5cm. Distinctive, small social 'finch' of grassland, savanna, rice fields. Often in flocks. Feeds on or near ground in seeding grasses, perches readily on fences, low trees. Sings from perch and in parachuting display flight. Nothing really similar in Belize; note stubby bill, face pattern, streaked back, male has yellow rump obvious in flight; cf. Lesser Goldfinch (p. 254). SOUNDS: Emphatic high *syiik!* or *psiep!* and short chipping phrases, *ssi-ssi chi* or *ss-siit*. Song slightly canary-like, a medley of sweet twittering trills mixed with *tik* and *speep* calls, can be prolonged. STATUS: Fairly common but local and somewhat nomadic. (Mexico to S America.)

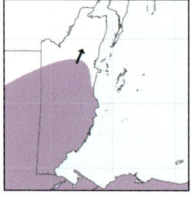

YELLOW-FACED GRASSQUIT *Tiaris olivaceus* 10–11cm. Small olive bird of weedy grassland and second growth, forest edge. Often with other small seed-eating birds in areas with seeding grasses. Sings from perch at low to mid-levels. Adult male striking and distinctive, female and imm. rather drab but note ghosting of male face pattern, conical bill. SOUNDS: High, thin, slightly sharp *sik*. Song a fairly rapid, high, ticking trill, 0.5–1 sec, at times in short series alternating trills of different length, pace, and pitch; easily passed off as an insect. STATUS: Uncommon to fairly common locally, especially in south; first recorded Belize around 1980 and still spreading north. (Mexico to nw. S America.)

CINNAMON-BELLIED
SALTATOR

imm.

BUFF-THROATED
SALTATOR

BLACK-HEADED
SALTATOR

variation

GRASSLAND
YELLOW-FINCH

female

male

juv.

YELLOW-FACED
GRASSQUIT

imm.
female

female

male

male

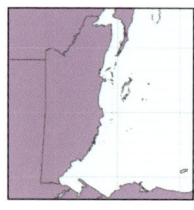

BLUE-BLACK GRASSQUIT *Volatinia jacarina* 10–11cm. Very small tanager of open grassy and weedy areas, rice fields, marshes. Often with other small seed-eating birds in areas with seeding grasses. Sings from low perch such as fence, weed stalk; often makes a short leap as it sings, showing off white shoulder tufts. Breeding male distinctive (can be seen year-round, depending on local conditions; mainly summer), other plumages told by small size, conical bill, dusky streaking on chest (can suggest small version of female Indigo Bunting, p. 260). Imm. male and nonbr. male have blue-black wings, variable dark blotching below. SOUNDS: High, sharp to slightly liquid *tsik*. Song a slurred, buzzy, slightly metallic or lisping *tzzzzz'u* or *tzssii'u*, every few secs, often as male leaps from perch. STATUS: Fairly common to common, especially in south. (Mexico to S America.)

SEEDEATERS, SEEDFINCHES (GENUS *SPOROPHILA*) (4 species). Small, stubby- and stout-billed tanagers formerly considered to be New World sparrows. Species often flock together, along with other seed-eating birds such as grassquits, *Passerina* buntings.

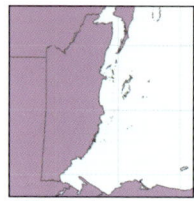

MORELET'S [WHITE-COLLARED] SEEDEATER *Sporophila (torqueola) morelleti* 10–11cm. Open grassy and weedy areas, roadsides, second growth, marshes, crop fields. Often with other small seed-eating birds in areas with seeding grasses. Sings from taller grass stalk, fence line, small shrub. Only seedeater in Belize with distinct pale wingbars, buff in fresh plumage, fading to whitish. Male variable: some relatively drab and olive-backed, plain below; others more boldly patterned black and whitish with broad blackish breast band. SOUNDS: Downslurred, slightly plaintive *chieh*, burrier nasal *chreh*. Song a varied warble of slurred whistles and rapid sweet chips, often ends with slurred buzzes; mainly 2–10 secs. STATUS: Fairly common to common; sporadic on Caye Caulker. (Mexico to w. Panama.)

SLATE-COLORED SEEDEATER *Sporophila schistacea* 11–11.5cm. Rare, stout-billed seedeater of humid forest edge and clearings, swampy wooded thickets, second growth, especially in areas with spiny bamboo. In pairs or small groups, sometimes with other seedeaters. Sings from mid-level perches, also in flight. Male distinctive, with stout yellowish bill, slaty plumage; on female note relatively plain plumage, stout bill with some pale at base. SOUNDS: Buzzy nasal *shih*, sometimes doubled; very high, slightly lisping *siik*. Song a varied, high, slightly buzzy and tinny twittering warble, often starts with an upslurred whistle, includes chipping trills; mainly 1–6 secs. STATUS: Rare and sporadic, mainly in the south. (Belize to S America.)

*BLACK [VARIABLE] SEEDEATER *Sporophila corvina* 11–11.5cm. Humid grassy second growth, forest edge, weedy thickets, roadsides; at times in adjacent seeding fields. Usually in pairs or small groups, sometimes with other small seed-eating birds. Male told from slightly larger Thick-billed Seedfinch by smaller, stubby bill, different song; female distinctive, plain olive-toned and often with male. SOUNDS: Nasal *chiyh* and downslurred *chieu*. Song a pleasant, slightly chanting warble, 2–8 secs; averages jerkier, burrier than Morelet's Seedeater. STATUS: Fairly common to common in south, less numerous and more local in north. (Mexico to w. Panama.)

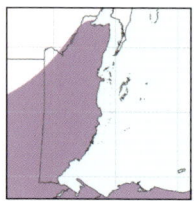

THICK-BILLED SEEDFINCH *Sporophila funerea* 11.5–12.5cm. Humid second growth, forest edge, marshes, adjacent seeding fields. Mostly in pairs, sometimes mixed with seedeaters and grassquits, but more often separate. Feeds inconspicuously in seeding grasses; sings from mid–upper levels in shrub or small tree. Cf. male Black Seedeater; female distinctive, often with male, but cf. Eastern Blue Bunting, usually in different habitats (p. 260). SOUNDS: Quiet nasal *chiyh*. Song a rich, slightly rambling, pleasant warble, at times with faster buzzy sections or run into a buzzy ending, mainly 5–25 secs; typically richer, more rambling, less jerky, and often more prolonged than seedeater songs. STATUS: Uncommon to fairly common locally. (Mexico to nw. S America.)

BLUE-BLACK GRASSQUIT

imm. male

female

breeding
male

MORELET'S SEEDEATER

female

males

variation

SLATE-COLORED SEEDEATER

male

imm. male

female

BLACK SEEDEATER

male

female

THICK-BILLED SEEDFINCH

female

male

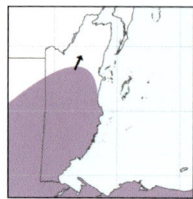

*COMMON BANANAQUIT *Coereba [flaveola] luteola* 9.5–10cm. Small, rather warbler-like tanager of humid forest and edge, adjacent second growth. Singles and pairs forage low to high in flowering trees; often sings from high in canopy. Probes and pierces flowers for nectar; also eats fruit, forages for insects. Nothing particularly similar in Belize (no overlap with Cozumel Bananaquit): note bold white eyebrow, pale grayish throat contrasting with yellow body, decurved bill, small white wing spot. **SOUNDS:** High, thin, sharp *seiit*, and thin, slightly lisping twitters. Song a rapid series of high, sibilant to buzzy notes run into a wiry trill or warble, 1–2 secs. **STATUS:** Fairly common to common, especially in Maya Mts; scarce and local in north, but range expanding. (Mexico to S America.)

*COZUMEL BANANAQUIT *Coereba [flaveola] caboti* 10.5–11cm. Northern large cayes, in gardens, forest edge, second growth. Habits similar to Common Bananaquit (no range overlap), but more of a garden bird, and at times in small groups. Bigger and brighter than Common, with whitish throat. Juv. duller overall with eyebrow and throat tinged yellowish. **SOUNDS:** High, thin, sharp *seit*, and thin, slightly lisping twitters. Song a high, thin, sibilant chipping or twittering trill, at times runs into a series of slightly lower, sweeter chips, 2–3.5 secs; less buzzy and jumbled than Common Bananaquit. **STATUS:** Fairly common to common on Ambergris and Caye Caulker. (Mexico to n. Belize.)

HONEYCREEPERS (3 species). Small colorful tanagers, often in pairs with mixed flocks, roving in forest canopy at fruiting and flowering trees. Calls are mainly high thin chips; songs rarely heard. Ages/ sexes differ; one species has seasonal change in male plumage.

GREEN HONEYCREEPER *Chlorophanes spiza* 13–14cm. Relatively large bulky honeycreeper of rainforest canopy and edge, adjacent second growth and semi-open areas with taller trees, fruiting shrubs. Usually found as singles or pairs, not groups, ranging in canopy of fruiting and flowering trees; often with mixed flocks. Appreciably larger than other honeycreepers; male stunning and distinctive, with black hood, banana-yellow bill; on female note yellowish bill, grayish legs. Juv. resembles female; male attains adult plumage over 1st year. **SOUNDS:** High sharp *tchiip*, rather warbler-like, may be repeated persistently; thin sharp *siip* mainly in flight. Rarely heard song a quiet buzzy twittering interspersed with short trills. **STATUS:** Uncommon to fairly common, especially in south. (Mexico to S America.)

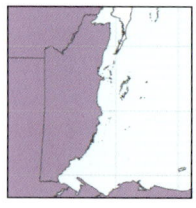

RED-LEGGED HONEYCREEPER *Cyanerpes cyaneus* 11–12cm. Humid forest and edge, adjacent open areas and second growth with taller trees, hedgerows, gardens. Often in small flocks, roving in canopy of flowering and fruiting trees, sometimes with other species. Note decurved bill, red legs; in flight, male shows bright sulphur-yellow underwings. Breeding male distinctive, but can look blackish high against bright light; female has dark eyestripe, weakly streaked underparts, cf. Green and Shining Honeycreepers. Nonbr. male like female but with black wings and tail. Juv. like female with duller pinkish legs; male has complete molt into plumage like adult nonbr. by early winter. **SOUNDS:** Buzzy, slightly overslurred mewing *meeah* or *meëihr*, suggesting a gnatcatcher; high, slightly nasal, rolled *srrip*; high thin *ssit* in flight. **STATUS:** Fairly common to common Mar–Aug; rare and sporadic in north in winter, when also less common in south. (Mexico to S America.)

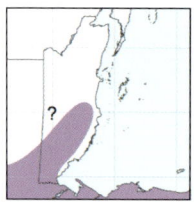

SHINING HONEYCREEPER *Cyanerpes lucidus* 10–10.5cm. Small, rather short-tailed honeycreeper of humid foothill and highland forest and edge, adjacent second growth with taller trees, fruiting shrubs. Singles and pairs range in canopy of fruiting and flowering trees; often with mixed flocks. Yellow legs usually conspicuous; also note more strongly arched bill, shorter tail than slightly larger Red-legged Honeycreeper. Male has distinctive black throat, female streaked bluish below. Juv. resembles female but breast streaking greenish, lacks distinct blue whisker; male attains adult plumage over 1st year. **SOUNDS:** High, thin, fairly sharp *chit* and thin twitters. **STATUS:** Fairly common above 600m in Maya Mts., rare and sporadic in adjacent foothills and lowlands. (Mexico to nw. Colombia.)

COMMON BANANAQUIT

juv.

COZUMEL BANANAQUIT

GREEN
HONEYCREEPER

male

female

RED-LEGGED HONEYCREEPER

female

male nonbr.
(Jun–Dec)

breeding male
(Dec–Jul)

female

male

SHINING
HONEYCREEPER

GENUS *TANGARA* (3 species). Classic *Tangara* are rather small, 'crown jewel' tanagers most diverse in foothills of the Andes; recent molecular work, however, shows that most tanagers traditionally placed in genus *Thraupis* (such as Blue-gray and Yellow-winged Tanagers) are embedded within *Tangara*. Usually in pairs or small groups in forest canopy, often at fruiting trees with mixed flocks.

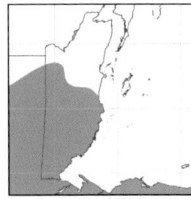

GOLDEN-HOODED TANAGER *Tangara (Stilpnia) larvata* 12.5–13cm. Handsome small tanager of humid forest and edge, adjacent clearings, second growth with taller trees. In pairs or small groups, mainly at mid–upper levels in fruiting and flowering trees; often with mixed flocks. Distinctive adult stunning in good light, with black body, golden hood, violet and turquoise highlights. Juv. usually with adult, shows ghosting of adult pattern. SOUNDS: Sharp, fairly hard chipping *chik* and clipped gruff *cheht*, which may be repeated steadily. Song a rapid dry trill often preceded by a few sharp chips or *cheht* notes, *cheht cheht ch ssiiiiiiiir*, 1–3 secs. STATUS: Fairly common, especially in south. (Mexico to nw. Ecuador.)

***BLUE-GRAY TANAGER** *Tangara (Thraupis) episcopus* 15.5–17.5cm. Familiar, often conspicuous tanager of open and semi-open areas with trees and bushes, towns, forest edge. In pairs or small groups, often in fruiting trees and associates readily with Yellow-winged Tanager; at times perches on high bare twigs and utility wires. Overall powder-blue plumage with brighter wings and dark beady eye distinctive. SOUNDS: Varied chips and lisping whistles, including high, slightly piercing, downslurred *ssiiu* and more nasal *sywee*, both of which may be given in flight. Song a high, slightly lisping to squeaky twittering warble, mostly 2–5 secs. STATUS: Fairly common to common in south, uncommon and more local in north; occasional on Ambergris Caye, mainly spring–summer. (Mexico to nw. Peru.)

YELLOW-WINGED TANAGER *Tangara (Thraupis) abbas* 17–19cm. Humid forest edge, semi-open areas with taller trees, gardens. Habits much like Blue-gray Tanager, and the 2 species associate together readily at fruiting trees. Beautiful purplish and lilac tones on head and breast often look drab grayish unless seen in good light, but note contrasting blackish wings with big yellow wing patch. SOUNDS: High, thin, overslurred or upslurred *ssiu* and *sweek*, both of which may be given in flight. Song a high, rapid, slightly spluttering or pulsating trill, often with 1–2 intro notes, *shee iiiiiiiiiiiir*, 2–3 secs, faster-paced, less dry and staccato than Golden-hooded Tanager song. STATUS: Fairly common to common in south, uncommon and local in north. (Mexico to Costa Rica.)

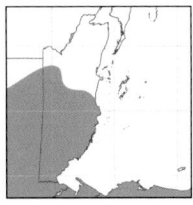

CRIMSON-COLLARED TANAGER *Ramphocelus sanguinolentus* 19–20cm. Handsome, distinctive tanager of humid forest edge, adjacent second growth. Mainly in pairs ranging low to high, especially at mid-levels; usually apart from mixed flocks. No similar species in Belize. Juv. duller overall, crimson areas of adult orange-red. SOUNDS: Piercing, high whistled *ssiiew*, and high thin *ssiip*, also in flight. Song a slightly jerky, unhurried medley of high, thin, sweet to slightly squeaky whistled notes, 1–2 notes/sec. STATUS: Fairly common, especially in south. (Mexico to w. Panama.)

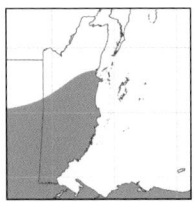

SCARLET-RUMPED (PASSERINI'S) TANAGER *Ramphocelus passerinii* 17–18cm. Distinctive tanager of humid second growth, forest edge, overgrown clearings, gardens. In pairs or small groups, mainly at low to mid-levels in leafy foliage, at times in association with other tanagers. Male unmistakable, velvet-black with flame-red rump often puffed out. Female distinctive, with silver-gray bill, grayish head, ochre-olive body. Juv. resembles female; adult male appearance attained over a few months in summer. SOUNDS: Dry, slightly nasal *cheht* or *chay*, often repeated in excited chatters; rough *shih* and lisping short chatters, such as *ssi-ssi cheh-chéh*, at times repeated; high buzzy *zzrit*. Song a variable short medley of (usually 2–5) slightly nasal whistled chirps, about 2 notes/sec, at times repeated steadily, such as *chiéh'i-wieh chiieh-wieh...*, or simply *chiéh-lii...*; might suggest a musical House Sparrow. STATUS: Fairly common to common, especially in south. (Mexico to w. Panama.)

GOLDEN-HOODED TANAGER

juv.

BLUE-GRAY TANAGER

juv.

YELLOW-WINGED TANAGER

juv.

CRIMSON-COLLARED TANAGER

SCARLET-RUMPED TANAGER

male

female

molting imm. male

BLACK-THROATED SHRIKE-TANAGER *Lanio aurantius* 20–21.5cm. Rather large arboreal tanager of rainforest. Pairs or family groups range mainly at mid–upper levels inside forest, usually with diverse mixed flocks for which it acts as a core member and sentinel. Perches upright and often still for long periods, watching for invertebrate prey to be flushed by flock before sallying out to snatch it; at times sounds an alarm call to distract other birds from prey. Relatively large size and behavior distinctive, and male plumage striking, but cf. orioles. Female best told by size, habits, stout hooked bill, also note tawny rump. Juv. resembles female but rustier-toned overall, mandible paler, grayish. **SOUNDS:** Slightly overslurred, clear whistled *wheeéu* and rich, downslurred, at times burry *teeeu*; both may be repeated loudly and steadily at intervals. Often prolonged series of rhythmic *ch'tu* calls may be sentinel alarm, usually preceded by short, slightly squeaky descending chatter; *ch'tu* calls sometimes run into short, slightly spluttering and clucking chatters, suggesting a *Turdus* thrush or Summer Tanager. Song infrequently given, a prolonged, slightly disjointed medley of squeaks, whistles, chips, splutters, and short warbled phrases. **STATUS:** Uncommon to fairly common, most numerous in south. (Mexico to n. Honduras.)

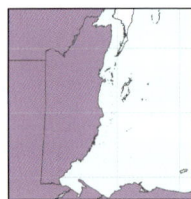

*GRAY-HEADED TANAGER *Eucometis penicillata* 16.5–18cm. Medium-size tanager of humid forest and edge, adjacent taller second growth. Pairs or family groups forage mainly at low to mid-levels in understory, often at army ant swarms, where can be shy and overlooked easily. Nothing really similar in Belize: note bright yellow underparts, sharply demarcated gray head, which often has a bushy crest, and habits. Juv. duller overall with olive head; soon attains adult plumage. **SOUNDS:** Hard, low, clipped *tuk*; high sharp *siip*, at times run into lisping twitters. Song a varied, usually rather fast-paced, rich, sharp warble; might suggest a euphonia; also a slightly jerky series of high, sharp, lisping chips. **STATUS:** Uncommon to fairly common. (Mexico to S America.)

NEW WORLD BLACKBIRDS (ICTERIDAE; 20+ SPECIES) Medium-size to large songbirds with pointed bills, strong legs and feet. Icterids include arboreal, fruit-eating orioles, oropendolas, and caciques; and largely terrestrial, seed-eating blackbirds, grackles, cowbirds. Ages/sexes similar or different, with males larger than females, sometimes strikingly so; attain adult appearance within 1st year.

*YELLOW-BILLED CACIQUE *Amblycercus holosericeus* 21–26cm, male> female. Skulking and furtive in second-growth thickets, forest edge, and understory tangles, often near water. Usually in pairs, moving deliberatively at low to mid-levels hidden in cover; often seen flying low across roads and trails, the second member of a pair following a few secs behind the first; sometimes attends army ant swarms. Note habits, pale ivory bill, staring pale yellow eyes. Juv. has dull eyes, sootier plumage, soon like adult. **SOUNDS:** Calls include short series of (usually 3–8) upslurred, slightly gruff, crowing to nasal notes, at times with slightly laughing or chuckling cadence overall, *shehr shehr…* or *yahnk yahnk…*, 4–5 notes/sec. Song a measured chant of rich, whistled, 2–3-syllable phrases, usually 2–8× with 1 phrase/1–1.5 secs, such as *heeu hih, heeu hih…* or *hoóee-hwee, hoóee-hwee…*, often overlapped by a slightly descending chattering rattle from mate, *hew chrrrrrrrrr*, 4–5 secs. **STATUS:** Fairly common to common. (Mexico to S America.)

Black-throated Shrike-Tanager

female

male

juv.

Gray-headed Tanager

Yellow-billed Cacique

juv.

MELODIOUS BLACKBIRD *Dives dives* 23–28cm, male>female. Common and conspicuous blackbird of open and semi-open habitats, forest edge, gardens. In pairs or small groups, feeding mainly on ground where walks confidently; associates readily with grackles. Often flicks tail up sharply, at times accompanied by whistles. Flight distinctive: slightly hesitant or jerky progression with rowing wingbeats, not steady and direct, groups loosely spaced. Best identified by shape, habits, voice; note medium size, sharply pointed bill, glossy black plumage ('velvet' plumage of face and neck apparent at close range). Cf. cowbirds. Juv. duller overall, soon like adult. **SOUNDS:** Sharp *weet!* or *piik!* sometimes repeated persistently. Varied, mostly 2-note, piercing whistles, often in short series, including *wh'chieuh* or *whi'chieh*, 2nd part downslurred; and a clipped, clear whistled *wh'dieeh*. **STATUS:** Common to fairly common; occasional on Ambergris Caye. (Mexico to w. Panama.)

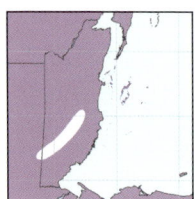

GREAT-TAILED GRACKLE *Quiscalus mexicanus* Male 40–49cm, female 32–39cm. Widespread, common, and noisy in open and semi-open habitats, especially near water. Feeds mainly on ground, striding confidently, at times in flocks of 100s that mix readily with cowbirds, less so with other blackbirds; roosts communally and noisily, often in town parks where many 1000s may gather. Flight fairly direct, with gentle low undulations, flocks loosely spaced and often strung out in lines. Distinctive in most of range, but in flight cf. Giant Cowbird. Juv. resembles female with dusky eyes, male larger than female; resembles adult by winter, but 1st-year male averages less glossy, shorter-tailed than adult. **SOUNDS:** Varied and noisy. Loud shrieks, whistles, clacks, and chatters. Bright, piercing, ascending whistled *wheeeeu* often ending emphatically; burry note followed by shriek. Flight call a gruff dry *chek* or *chuk*. **STATUS:** Common and widespread except in forested areas; rare wanderer to Mountain Pine Ridge. (Mexico and sw. US to nw. S America.)

COWBIRDS (GENUS *MOLOTHRUS*)

(3+ species). Medium-size to large blackbirds of open and semi-open country, often found around livestock. Ages/sexes differ slightly to distinctly, attain adult appearance in a few months; males slightly to distinctly larger than females. Brood parasites of various species; hence begging young cowbirds are found singly, being fed by sundry species but not by cowbirds.

BRONZED COWBIRD *Molothrus aeneus* 19–22cm, male>female. Ranchland (often around livestock), open country with scattered trees, hedgerows, parks and playing fields in towns, villages; more often woodland and forest edge in breeding season. Singles and small groups in breeding season; flocks to 100+ birds at other times. Feeds on ground, tail often slightly cocked; roosts in trees, often with grackles. Perches readily in trees and on phone wires; male in display expands neck ruff and hovers in front of female, singing. Flight strong and direct, often in fairly tight, slightly undulating flocks. Best identified by stout, deep-based bill, reddish eyes (fiery-red on male, duller on female and imm.); adult male wings have glossy bluish sheen. Juv dark sooty brown to dull blackish overall, sometimes with fine pale wingbars, faint pale streaking on belly, like dull adult by winter. **SOUNDS:** High, thin, tinny whining trills and wheezy squeaks, at times suggesting European Starling *Sturnus vulgaris*. Spluttering, chattering rattle, 1–2 secs (mainly female). Song a varied series of high whining whistles, at times mixed with short quiet gurgles. **STATUS:** Fairly common to common in north, less numerous and more local in south; more widespread in winter. (Mexico and sw. US to Panama.)

SHINY COWBIRD *Molothrus bonariensis* Male 18–19cm, female 16.5–17.5cm. Vagrant. Habitat and habits much like Bronzed Cowbird, with which Shiny may be found. Slightly smaller than Bronzed, with shallower, more pointed bill. Male has glossy purplish-blue head and body, dark eyes; also cf. Melodious Blackbird. Female has plain pale gray-brown underparts. **SOUNDS:** Male gives high thin *seeih*; both sexes (mainly female) give clucking or bubbling rattle, 1–2 secs, averaging lower than Bronzed Cowbird. Song a high, slightly metallic to fairly sweet warble, usually overall descending, 1.5–4 secs, often preceded by series of (usually 4–8) low, wet gulping grunts that can be given separately. **STATUS:** Very rare, with 2 records (Aug–Oct), first in 2007. (Caribbean and S America.)

MELODIOUS BLACKBIRD

males

GREAT-TAILED GRACKLE

female

juv.

female

BRONZED COWBIRD

male

SHINY COWBIRD

female

male

GIANT COWBIRD *Molothrus oryzivorus* Male 32–36cm, female 28–30.5cm. Large cowbird of semi-open country, ranchland with taller trees, forest edge and clearings; brood parasite of oropendolas. Mostly seen in flight, often fairly high overhead, as singles or small loose groups: note distinctive, strong, flap-flap-flap-glide progression, unlike steadier wingbeats of Great-tailed Grackle. Feeds on ground around livestock, along riverbanks, but arboreal at oropendola colonies. Much larger than other cowbirds. In flight note pointed wings, thick neck accentuating small-headed look, squared-off tail; cf. Great-tailed Grackle. Juv. has pale bill, pale to brownish eyes; soon resembles duller version of adult. **SOUNDS:** Sharp clucks and short chatters mainly in interactions. Song a short, slightly jerky or discordant series of harsh, semi-metallic whistles. **STATUS:** Uncommon to fairly common, especially in south. (Mexico to S America.)

OROPENDOLAS (GENUS *PSAROCOLIUS*) (2 species). Very large, social, arboreal 'blackbirds' with stout pointed bills, bright yellow tail sides. Ages differ slightly, soon attain adult appearance; sexes similar in plumage but males much larger than females. Colonies of woven pendulous nests conspicuous, often attended by brood-parasite Giant Cowbirds.

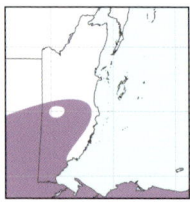

CHESTNUT-HEADED OROPENDOLA *Psarocolius wagleri* Male 34–38cm, female 25–29cm. Medium-size dark oropendola of humid forest and edge; colonies typically at edges, often in isolated trees. Flight rather quick with deep swooping wingbeats that make hollow rushing sound, unlike unhurried, measured flight of Montezuma Oropendola. Usually in small groups, feeding in fruiting and flowering trees where clambers actively. Juv. has duller eyes and bill, back and wings lack sheen of adult. **SOUNDS:** Varied throaty clucks and low chatters, often more guttural than Montezuma Oropendola. Low, slightly gurgling *chuk-uk-luk*; nasal *kyah*; slightly nasal clucking *chéuh*. Song a deep gurgled *w'shLOK*, and variations, repeated. **STATUS:** Uncommon to fairly common locally, especially in foothills. (Mexico to w. Ecuador.)

MONTEZUMA OROPENDOLA *Psarocolius montezuma* Male 45–51cm, female 37–41cm. Distinctive, very large (male much larger than female) and colorful inhabitant of humid forest edge, semi-open country with hedgerows, scattered trees; colonies typically in isolated trees. Flight unhurried and steady, usually at tree-top height and often across open areas, with measured rowing wingbeats; often in strung-out flocks that may suggest Brown Jay. Feeds mainly at mid–upper levels in fruiting and flowering trees, also on ground at times, singly or in small groups. In display, male swings upside-down on perch, spreading wings and fanning tail. Juv. duller overall, underparts sooty blackish. **SOUNDS:** Squeaky *woik*; sneezy *rrúh-rrúh*; and varied gruff hollow clucking series; low *chuk* often given in flight. Song memorable: a bizarre, hollow gurgling and gobbling crescendo, ending abruptly, 1–1.5 secs; likened to pouring wine from an upended bottle and sometimes followed by a thin upslurred whistle; raspy wing rattles in display. **STATUS:** Fairly common to common. (Mexico to Panama.)

cf. Great-tailed Grackle

juv.
(May–Aug)

female

male

GIANT COWBIRD

CHESTNUT-HEADED
OROPENDOLA

MONTEZUMA
OROPENDOLA

AMERICAN ORIOLES (GENUS *ICTERUS*) (9+ species). Colorful arboreal 'blackbirds,' many with complex age/sex plumages. For ID note overall size and structure, especially bill size and shape, plus wing patterns; many potentially similar species also separate by distribution. Most species feed low to high in fruiting and flowering trees, where different species associate readily when feeding. Calls notably varied.

BLACK-COWLED ORIOLE *Icterus prosthemelas* 18.5–21cm. Relatively small, yellow-and-black oriole of humid forest and edge, clearings and semi-open areas with hedgerows, scattered trees, gardens. Often in pairs, perching atop leafy trees in early morning; associates readily with other orioles at flowering trees. Plumage can suggest several other oriole species but note slender, slightly decurved bill. Larger Yellow-backed Oriole has smaller black bib, straight pointed bill, lacks yellow shoulder; imm. Yellow-tailed Oriole larger and bigger billed with less black in face, yellow tail sides. SOUNDS: Not a particularly vocal species. Calls include a plaintive nasal *tchieh* and downslurred nasal *nyeh*; both may be repeated steadily. Song a pleasant, slightly jerky rich warble, 1–2 secs. STATUS: Fairly common. (Mexico to w. Panama.)

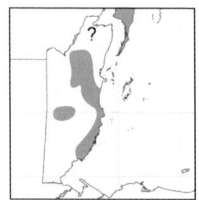

YELLOW-BACKED ORIOLE *Icterus chrysater* 22–24cm. Handsome oriole of pine savanna, locally also semi-deciduous forest and edge, adjacent clearings. In pairs or small groups at flowering and fruiting trees, poking into bromeliads; at times with mixed flocks including jays, other orioles. Note straight pointed bill, boldly patterned plumage with yellow back, relatively extensive black bib. Male brighter than female; imm. duller overall but still distinctive. Female and imm. Black-cowled Oriole often mistaken for Yellow-backed. SOUNDS: Rough nasal *ehrr* or *yehnk*, often in short series; plaintive downslurred *chuu*. Song a varied medley of sad rich whistles, often unhurried and rather hesitant, recalling Audubon's Oriole *I. graduacauda* of Mexico; also faster versions, at times with repeated rhythmic phrases and nasal calls included. STATUS: Fairly common to uncommon but often local. (Mexico to nw. S America.)

YELLOW-TAILED ORIOLE *Icterus mesomelas* 20.5–23.5cm. Handsome oriole of humid second-growth thickets and forest edge, often along rivers and in damp areas. Usually in pairs or small groups, often rather skulking at low to mid-levels in tangles; at other times in more open flowering trees and shrubs with other orioles. Adult stunning and flashy; imm. duller but also distinctive, note stout, slightly decurved bill, yellow tail sides (can be all but absent during tail molt). Tail pattern often striking as birds dive into cover, cf. Mayan Green Jay. SOUNDS: Rich twangy *cheuk* and downslurred *cheu*, at times repeated steadily. Song a varied, rhythmic to slightly rollicking repetition (usually 4–20×) of rich, 2–7-note whistled phrases, such as *tch wee-choo weeep, tch wee-choo weeep…*; sometimes in duet. STATUS: Fairly common to uncommon. (Mexico to nw. Peru.)

BLACK-COWLED ORIOLE

juv.
(May–Aug)

female

male

nest pouch slung
under palm frond

YELLOW-BACKED ORIOLE

juv.
(May–Aug)

female

male

YELLOW-TAILED ORIOLE

1st-year

juv.
(May–Aug)

adult

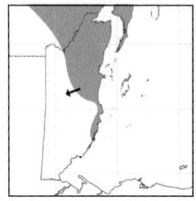

HOODED ORIOLE *Icterus cucullatus* 18.5–20cm. Medium-size, slender, long-tailed oriole with slender decurved bill. Varied semi-open and wooded habitats from beach scrub to humid forest edge; almost invariably near palms and palmettos when nesting. Distinctive, with slender decurved bill, long tail, orange plumage tones; female lacks black bib. Juv. like female but with broader, buffier wingbars, male soon attains black on bib. Cf. very local Orange Oriole, larger and bulkier Altamira Oriole. SOUNDS: Nasal, slightly metallic rising *wiehn?* Short gruff chatters, and hard dry *chek*. Song a fairly fast-paced, scratchy and jerky warble, 1–3 secs, often mixed with harsh *chek* notes. STATUS: Fairly common to common in north, including larger cayes; uncommon to scarce and local in south; spreading locally. (Breeds sw. US and Mexico, winters Mexico.)

ORANGE ORIOLE *Icterus auratus* 18–19cm. Very local in northeast. Medium-size, rather chunky oriole of brushy woodland, forest edge, semi-open areas with scattered trees. Often in pairs; associates readily with Hooded and Altamira Orioles at flowering and fruiting trees. Male distinctive, with orange back, smaller black mask than male Hooded. Female and imm. best told from imm. male Hooded by stockier build, straight pointed bill, smaller black mask, bolder white edging at base of primaries. SOUNDS: Varied calls, some very similar to Altamira Oriole. Fairly hard nasal *nyeh*; downslurred rich *cheu*; rapid dry chatter, sharper and more rattling than Hooded Oriole; slightly nasal, drawn-out *wheet*; and plaintive *hui?* Song a varied slow medley of clear to slightly plaintive whistles, may recall Altamira Oriole but faster-paced, quieter; also a rapid ringing series of (mostly 5–16) slightly nasal whistles, *chiehchieh…*, 6–8 notes/sec, almost too fast to count. STATUS: Uncommon and local in n. Corozal, including Ambergris, where most frequent in winter. (Mexico to Belize.)

ALTAMIRA ORIOLE *Icterus gularis* 23–25.5cm. Largest oriole, big and bulky with deep-based bill. Varied wooded habitats, forest edge, semi-open areas with scattered trees, hedgerows. Often in pairs or small groups; associates readily with other orioles at flowering and fruiting trees. Large size and deep, mostly dark bill distinctive, also note orange shoulder, single bold white wingbar. 1st-year paler overall with olive back, best identified by size, bill shape. SOUNDS: Varied calls include a harsh nasal *nyeh* or *yehnk*, often doubled or in steady scolding series; bright, slightly ringing *chiu*; sharp *peen*; and varied rich whistled notes, often in hesitant series, *chu*, *wee-chu* and *chee'chu*, repeated; lacks dry chatter call of other orange orioles. Song an unhurried, often slightly jerky or staccato medley of loud rich whistles, sometimes with a burst of 1 note repeated in fairly rapid series. STATUS: Uncommon to fairly common in north; occasional on Ambergris Caye, mainly in winter; recent record from s. coast may mirror colonization from Guatemala by Spot-breasted Oriole. (Mexico and s. Texas to nw. Nicaragua.)

SPOT-BREASTED ORIOLE *Icterus pectoralis* 21–23cm. Recent colonist of s. coast. A fairly large oriole of semi-open areas with scattered trees, forest edge, gardens. Often in pairs. Adult distinctive, with black back, variable breast spotting, white wing panel. Also note slightly decurved bill, lack of white wingbars; cf. Altamira Oriole. 1st-year best told by wing pattern, bill shape; often has some black spots on breast sides. SOUNDS: Nasal *nyeh*, often repeated a few times; clipped squeaky *tchiu*; relatively slow-paced, harsh staccato chatter, *cheh-cheh…*, often fairly soft. Song a pleasing, unhurried warble or slow chant of rich slurred whistles, often some phrases repeated with rhythmic caroling cadence, at times in duet; more flowing and rhythmic, less staccato than Altamira Oriole song. STATUS: Scarce and local (but spreading) in south coastal areas, where first recorded 2007. (Mexico to nw. Costa Rica.)

1st-year male

female

male

HOODED ORIOLE

female

ORANGE ORIOLE

male

ALTAMIRA ORIOLE

1st-year

juv. (Jul–Sep)

adult

1st-year

juv. (Jul–Sep)

adult

SPOT-BREASTED ORIOLE

ORCHARD ORIOLE *Icterus spurius* 15–17cm. Small migrant oriole of hedgerows, gardens, overgrown weedy fields, marshes, forest edge. Often in small groups, feeding low to high in flowering trees and bushes. Small size and small, slightly decurved bill distinctive. Chestnut body plumage of adult male unique among orioles; female and imm. greenish yellow with distinct white wingbars. 1st-year male resembles female with variable black bib, sometimes patches of adult color in spring–summer. SOUNDS: Low gruff *chuk* and gruff chattering *chuh-chuh…*, lower and more spluttering, less dry and staccato than Hooded and Baltimore Oriole chatters; quiet nasal *huih*. STATUS: Fairly common to common Sep–Mar; more widespread and locally numerous in migration, mid-Jul to Sep, Mar to mid-May. (Breeds N America to Mexico, winters Mexico to nw. S America.)

BALTIMORE ORIOLE *Icterus galbula* 18–20.5cm. Winter migrant to forest edge, second growth, semi-open areas with hedgerows, gardens. Often in small groups, moving in canopy of flowering and fruiting trees; mixes readily with other orioles. Adult male striking and distinctive, with black head and back, orange body; female and imm. variable, but always with straight pointed bill, orange plumage tones, bold white wingbars; some adult females and imm. males have extensive black mottling on head and back. SOUNDS: Gruff staccato chatter; occasional mellow whistles, mainly in spring. STATUS: Fairly common to common Oct–Apr; more widespread and locally numerous in migration, Sep–Oct, Apr to mid-May. (Breeds e. N America, winters Mexico to nw. S America.)

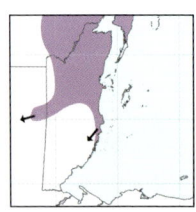

RED-WINGED BLACKBIRD *Agelaius phoeniceus* Male 20–22cm, female 17–18cm. Breeds colonially in fresh marshes with reedbeds, tall weedy growth along rivers; ranges widely to feed in farmland and open country. Sings from low perch or in short display flight; feeds mainly on ground, often in flocks outside breeding season. No similar species in Belize, but note male's red shoulder can be covered at rest. Juv. resembles female but paler overall; 1st-year male variably intermediate between female and male. SOUNDS: Gruff clipped *chek* or *chwek* in flight; slightly metallic high *dink*. Song a short, strangled, slightly metallic trill with a warbled introduction. STATUS: Fairly common to common locally in north, scarce and local in south but spreading; rare and sporadic s. of mapped range. (N America to Costa Rica.)

BOBOLINK *Dolichonyx oryzivorus* 15–17cm. Scarce transient icterid of rough grassland, crop fields, marshes. Typically feeds in grasses and rarely seen unless flushed, when may perch atop a bush calling. Usually singles, but can occur in small flocks; flight strong and direct. Imm. and nonbr. adults similar, with unstreaked rich buffy breast, bold head stripes, stout pointed bill, spiky tail. Breeding male distinctive, with black face and underparts, buff cowl; fresh plumage veiled buff; breeding female like nonbr. SOUNDS: Semi-metallic, clipped nasal *iihnk* mainly in flight; gruff clipped *chuhk!* STATUS: Scarce Sep–Oct, mid-Apr to early Jun, mainly on cayes; very rare on mainland. (Breeds N America, winters S America.)

EASTERN MEADOWLARK *Sturnella magna* 19–24cm, male>female. Distinctive, rather stocky icterid of grassy savanna, farmland, other open grassy areas. Feeds on ground, where walks strongly. Sings from ground or low perch. Flight strong, flushing explosively from close range and flying with fairly rapid stiff wingbeats and flat-winged glides before dropping back to cover, when tail spreads to show off white sides. No similar species in Belize. Female averages smaller, duller than male; juv. duller and paler than adult with streaky necklace; all fresh plumages heavily veiled buff on underparts, wearing away to reveal golden yellow. SOUNDS: Raspy *zzzrt*; spluttering rattle (suggests female cowbird); nasal *sweink*, mainly in flight. Song a simple arrangement of downslurred plaintive whistles, 1–1.5 secs, such as *see syeée teu-eh*. STATUS: Fairly common to common but local; range expanding with deforestation. (N America to Panama.)

ORCHARD ORIOLE

imm. male

female

male

BALTIMORE ORIOLE

imm.

female

male

RED-WINGED BLACKBIRD

female

male

BOBOLINK

female
(imm./nonbr.
male similar)

breeding male
(Apr–May)

juv.

male

EASTERN MEADOWLARK

APPENDIX A

Rare Migrants and Vagrants

Here we list 81 species reported from Belize as rare migrants and vagrants, plus one extirpated breeding species. Some of these are likely of more regular occurrence than presently known and, if encountered, any of these species should be documented carefully. In a few cases, plate layout allowed some of these species to be incorporated in the main body of the book; these are mainly species that appear similar to more common and regularly occurring species (such as White-faced Ibis vs. Glossy Ibis) or species that might be encountered by anyone out in the field and for which inclusion may help clarify status (such as Cayenne Lapwing, Shiny Cowbird).

^{V} Very rare, sporadic, or potentially overlooked; not usually recorded annually but could be encountered on occasion, such as Dunlin, American Pipit, Hermit Warbler; plus one former breeding species (Atlantic Black Noddy), which might still occur on occasion as a vagrant.

^{X} Exceptional; species with at most 3 records and not to be expected, such as Black-tailed Gull, Yellow-naped Amazon, Prevost's Ground Sparrow.

(Parentheses indicate no substantiated records or reports of presumed wild birds since 2000, such as Mallard, Atlantic Black Noddy, Short-eared Owl.)

[Brackets indicate species reported in the literature or eBird, but without convincing documentation or of uncertain provenance; here considered hypothetical, such as Great Black-backed Gull, Green Parakeet, Virginia's Warbler.]

^{V}Snow Goose *Anser caerulescens*
(^{X}Greater White-fronted Goose *Anser albifrons*)
^{X}Canada Goose *Branta canadensis*
(^{X}Mallard *Anas platyrhynchos*)
^{X}Gadwall *Mareca strepera*
^{V}Redhead *Athya americana*
^{V}Greater Scaup *Aythya marila*
^{X}Hooded Merganser *Lophodytes cucullatus*
^{V}Red-breasted Merganser *Mergus merganser*
^{V}Ruddy Duck *Oxyura jamaicensis*

^{X}Cory's Shearwater *Calonectris [diomedea] borealis*
^{X}Great Shearwater *Ardena gravis*
(^{X}Manx Shearwater *Puffinus puffinus*)
^{V}Audubon's Shearwater *Puffinus lherminieri*
^{X}Atlantic Yellow-nosed Albatross *Thalassarche chlororhynchos*
^{V}Wilson's Storm-Petrel *Oceanites oceanicus*
^{V}Band-rumped Storm-Petrel *Thalobata (Hydrobates) castro*
^{V}Catesby's [White-tailed] Tropicbird *Phaethon [lepturus] catesbyi*
^{X}Red-billed Tropicbird *Phaethon aethereus*
(^{X}Great Skua *Catharacta skua*)
^{V}Bonaparte's Gull *Chroicocephalus philadelphia*
^{X}Black-legged Kittiwake *Rissa tridactyla*
(^{X}Black-tailed Gull *Larus crassirostris*)
^{X}Thayer's [Iceland] Gull *Larus [glaucoides] thayeri*

[Great Black-backed Gull *Larus marinus*]
(^{V}Atlantic Black Noddy *Anous [minutus] americanus*)
^{V}Arctic Tern *Sterna paradisaea*

^{X}Cayenne Lapwing *Vanellus cayennensis*
^{V}Snowy Plover *Charadrius nivosus*
^{V}Dunlin *Calidris alpina*
^{V}Baird's Sandpiper *Calidris bairdii*
^{V}Ruff *Calidris pugnax*
^{V}Hudsonian Godwit *Limosa haemastica*
^{V}Red-necked Phalarope *Phalaropus lobatus*
^{V}American Flamingo *Phoenicopterus ruber*
^{V}White-faced Ibis *Plegadis falcinellus*
(^{X}Scarlet Ibis *Eudocimus ruber*)

^{X}Bald Eagle *Haliaeetus leucocephalus*
(^{X}Short-eared Owl *Asio otus*)
^{V}Burrowing Owl *Athene cunicularia*
[Zenaida Dove *Zenaida aurita*]
^{X}Dark-billed Cuckoo *Coccyzus melacoryphus*
[Green Parakeet *Psittacara holochlora*]
^{X}Yellow-naped Amazon *Amazona auropalliata*
[Violet-headed Hummingbird *Klais guimeti*]
[Black Swift *Cypseloides niger*]
^{X}Brown-banded [Brown-chested] Martin *Phaeoprogne (Progne) [tapera] fusca*

[Southern Rough-winged Swallow *Stelgidopteryx ruficollis*]

[Sinaloa/Caribbean Martin *Progne sinaloae/ dominicensis*]

^VViolet-green Swallow *Tachycineta thalassina*

^VWestern Pewee *Contopus sordidulus*

[Eastern Phoebe *Sayornis phoebe*]

^XCassin's Kingbird *Tyrannus vociferans*

^VWestern Kingbird *Tyrannus verticalis*

^XNorthern Mockingbird *Mimus polyglottos*

^VAmerican Robin *Turdus migratorius*

^VHermit Thrush *Catharus guttatus*

[Bicknell's [Gray-cheeked] Thrush *Catharus [minimus] bicknelli*]

^XRuby-crowned Kinglet *Regulus calendula*

[Bell's Vireo *Vireo bellii*]

^VEastern Warbling Vireo *Vireo gilvus*

^XOrange-crowned Warbler *Leiothlypis celata*

^VMacGillivray's Warbler *Geothlypis tolmei*

^VConnecticut Warbler *Oporornis agilis*

[Virginia's Warbler *Oreothlypis virginiae*]

^XGolden-cheeked Warbler *Setophaga chrysoparia*

^XTownsend's Warbler *Setophaga townsendi*

^VHermit Warbler *Setophaga occidentalis*

[Black-throated Gray Warbler *Setophaga nigrescens*]

^VAudubon's Warbler *Setophaga [coronata] auduboni*

^VAmerican Pipit *Anthus rubescens*

[Lazuli Bunting *Passerina amoena*]

[Black-headed Grosbeak *Pheucticus melanocephala*]

^XPrevost's (White-faced) Ground Sparrow *Melozone biarcuata*

^XVesper Sparrow *Pooecetes gramineus*

^VWhite-crowned Sparrow *Zonotrichia leucophrys*

^XWhite-throated Sparrow *Zonotrichia albicollis*

^XBrown-headed Cowbird *Molothus ater*

^XShiny Cowbird *Molothrus bonariensis*

[Bullock's Oriole *Icterus bullocki*]

^XYellow-headed Blackbird *Xanthocephalus xanthocephalus*

APPENDIX B

Taxonomic Notes

Here we summarize reasons for treating various taxa as potentially separate species and also mention other possible taxonomic options for some species; these are marked in the species accounts by an asterisk (*) preceding the name and by use of brackets, for example, American Great Egret *Ardea [alba] egretta* or Aztec [Olive-throated] Parakeet *Eupsittula [nana] aztec*, as explained on pp. 7–8. The term 'group' is used for one or more subspecies that form a potential species; group names refer to the chronologically first-named taxon in a group.

In some cases, differences (as in plumage, morphology, vocalizations, ecology) are so clear that there is no question different species are involved (e.g., Cayenne Lapwing, Yucatan Dove); in other cases, more work is needed, and our mention of differences is intended to draw attention in the hope of promoting careful study. Our baseline taxonomy is that of IOC (Gill et al. 2021), except for oceanic birds, for which we follow Howell & Zufelt (2019), namely for the Brown Booby, White-tailed Tropicbird, Bridled Tern, and Brown Noddy complexes. In cases where our opinions differ from IOC (several IOC splits we find unconvincing), the following notes summarize our reasons.

We have attempted to consider taxonomy within Middle America and between Middle America and at least adjacent South America. Taxonomy across all of South America, however, is beyond our purview, although in some cases it is clear that Central American taxa are distinct. Thus we may break species into Middle American taxa and 'all the rest,' including potentially multiple species in South America that we lacked the resources to investigate. Potential species-level differences for species in which all groups occur in Belize are summarized in the species accounts and not expanded on below, namely for Olive-sided Flycatcher, Swainson's Thrush, and Wilson's Warbler.

Depending on available data and philosophy, our opinions vary for different taxa, grading from "suggest that A and B may represent separate species" to "indicate that A and B are separate species." Probably every one of these cases represents a species, or lineage, according to some species concept (Howell 2021), as do others not mentioned here. We only include taxa for which we have biological support for possible species status (rather than simply uninformed molecular data), but undoubtedly we have still underestimated the number of candidates.

BLACK-BELLIED WHISTLING DUCK *Dendrocygna autumnalis* (p. 26). Differences in plumage and voice suggest that the *fulgens* group (N America to w. Panama) and *autumnalis* group (Panama to S America) may represent separate species. Plumage appears to intergrade from s. Costa Rica to cen. Panama, but study is needed.

SMITHSONIAN GULL *Larus [argentatus] smithsonianus* (p. 40). Molecular work and juvenile plumages indicate that N American populations of 'Herring Gull' are distinct from European populations and are more closely related to N Pacific taxa such as Glaucous-winged Gull *L. glaucescens*, as summarized by Olson & Banks (2007). As proposed by those authors, the English name 'Smithsonian Gull' removes misleading association with 'the Herring Gull' of Europe and we prefer this option over American Herring Gull, used by other authors such as IOC.

SANDWICH TERN *Thalasseus sandvicensis* (p. 42). Molecular work suggests that 'American Sandwich Tern' is more closely related to Elegant Tern than to 'European Sandwich Tern,' and some authors (including IOC) split N American populations as Cabot's Tern *T. acuflavidus*. We find the molecular studies weak, and unconvincing as evidence for biological species status; either way, retaining simply 'Sandwich Tern' for one of the taxa (as done for the European birds) is ambiguous and needlessly confusing (also see Howell 2021).

AMERICAN BLACK TERN *Chlidonias [niger] surinamensis* (p. 46). Differences in plumage and provisionally in voice suggest that the *surinamensis* group (New World) and *niger* group (Old World; Eurasian Black Tern) represent separate species.

BLACK SKIMMER *Rynchops niger* (p. 46). Differences in plumage and ecology between the widespread *niger* group (coastal Americas; American Skimmer) and *cinerascens* group (interior S America; Amazonian Skimmer) suggest cryptic species may be involved.

CAYENNE [SOUTHERN] LAPWING *Vanellus [chilensis] cayennensis* (p. 50). Differences in morphology, plumage, and vocalizations indicate that the widespread *cayennensis* group (S America, spreading to Central America) and allopatric *chilensis* group (Chile and s. Argentina; Chilean Lapwing) are best considered separate species, as done by Howell & Schmitt (2018).

WHIMBREL *Numenius phaeopus* (p. 54). Species status has been argued for New World populations (*hudsonicus* group, Hudsonian Whimbrel), based largely on genetic data (Sangster et al. 2011); although Hudsonian Whimbrel is split by IOC, in our view the case remains unproven from a biological perspective.

WILLET *Tringa semipalmata* (p. 56). Differences in ecology, morphology, plumage, and voice (supported by molecular evidence; Oswald et al. 2016) indicate that the *inornata* group (breeding interior w. N America; Western Willet) and *semipalmata* group (breeding coastal e. N America; Eastern Willet) are probably best treated as separate species.

AMERICAN GREAT EGRET *Ardea [alba] egretta* (p. 64). Differences in morphology, seasonal bare-part coloration, and provisionally in voice indicate that the *egretta* group (Americas) is best treated as a species distinct from Old World populations.

GREAT WHITE HERON *Ardea [herodias] occidentalis* (p. 64). Differences in morphology, plumage, and ecology suggest that the *occidentalis* group (Caribbean region) and *herodias* group (N America; Great Blue Heron) are probably best treated as separate species.

BOAT-BILLED HERON *Cochlearius cochlearius* (p. 68). Differences in plumage coloration, crest length of breeding birds, and provisionally in voice (averages higher, more chuckling in Northern), suggest that the *zeledoni* group (Middle America; Northern Boat-billed Heron) and *cochlearius* group (S America; Southern Boat-billed Heron) may represent separate species.

LIMPKIN *Aramus guarauna* (p. 74). Differences in plumage and provisionally in voice suggest that the *pictus* group (s. to w. Panama; Northern Limpkin) and *guarauna* group (cen. Panama to S America; Southern Limpkin) may represent separate species.

NORTHERN BLACK RAIL *Laterallus jamaicensis* (p. 78). Differences in voice and plumage indicate that the widespread *jamaicensis* group (Americas) and *salinasi* group (Chile and w. Peru; Chilean Black Rail) are best treated as separate species, as done by Howell & Schmitt (2018).

NORTHERN TURKEY VULTURE *Cathartes aura* (p. 90). Differences in morphology, plumage, and bare-part coloration indicate that the Turkey Vulture complex comprises at least 3 groups and at least 2 species: Northern Turkey Vulture (Mexico to Costa Rica); Tropical Turkey Vulture *C. [a.] ruficollis* (Panama to tropical S America); and the distinctive Austral Turkey Vulture *C. [a.] jota* (Andes and s. S America, including *falklandicus*).

NORTHERN WHITE HAWK *Pseudastur [albicollis] ghiesbreghti* (p. 104). Differences in voice and morphology indicate that the *ghiesbreghti* group (Mexico to nw. S America) and *superciliaris* group (S America, e. of the Andes; Southern White Hawk) are best treated as separate species; also see Lerner et al. (2008).

OSPREY *Pandion haliaetus* (p. 92). Wink et al. (2004) argued that all 3 Osprey taxa they studied (of 4 worldwide) warranted species status, based purely on mitochondrial DNA lineages, whereas Monti et al. (2015) provided a conflicting molecular interpretation based again purely on mitochondrial DNA. Although IOC and others have adopted a two-way split (Western Osprey *P. [h.] haliaetus* vs. Eastern Osprey *P. [h.] cristatus*), we see no biological evidence that offers compelling support for this; a 4-way species split seems equally valid, or all taxa might be maintained as a single species.

MESOAMERICAN [VERMICULATED] SCREECH OWL *Megascops guatemalae* (p. 124). On the basis of morphological and vocal differences comparable to those between other species of screech owls, we recognize 5 species of 'Vermiculated Screech Owl' in Middle America, including the *guatemalae* group (e. Mexico to Nicaragua), as distinct from West Mexican Screech Owl *M. [g.] hastatus* of w. Mexico, both of which are combined into Middle American Screech Owl by IOC.

FERRUGINOUS PYGMY OWL *Glaucidium brasilianum* (p. 124). Vocal and slight morphological and plumage differences suggest that the *ridgwayi* group (Mexico to nw. Colombia; Ridgway's Pygmy Owl) and *brasilianum* group (S America, e. of Andes; Ferruginous Pygmy Owl) may best be treated as separate species.

NORTHERN MOTTLED OWL *Strix virgata* (p. 120). Differences in voice and morphology indicate that the *virgata* group (Mexico to nw. Peru) and *superciliaris* group (S America, e. of the Andes; Amazonian Mottled Owl) are best treated as separate species. Also see Sanchez et al. (2012).

BARN OWL *Tyto alba* (p. 122). The Barn Owl *Tyto alba* complex comprises 3 genetic lineages that are treated differently by different authors. American Barn Owl *T. [a.] furcata* is treated as a species by IOC, but critical biological analysis of the situation could help resolve this issue.

LAWRENCE'S [WHITE-TIPPED] DOVE *Leptotila [verreauxi] fulviventris* (p. 136). Differences in song and bare-part coloration indicate that the *fulviventris* group (Mexico to w. Nicaragua) and *verreauxi* group (Nicaragua to S America; Verreaux's Dove) are best treated as separate species, as they were once treated before being lumped with no rationale.

YUCATAN [CARIBBEAN] DOVE *Leptotila [jamaicensis] gaumeri* (p. 136). Differences in song, plumage, and morphology indicate that the *gaumeri* group (Yucatan Peninsula) and *jamaicensis* group (Jamaica, Cayman Is.; Caribbean Dove) are best treated as separate species.

COMMON SQUIRREL CUCKOO *Piaya cayana* (p. 142). Differences in plumage and voice indicate that the *mexicana* group (w. Mexico; Mexican Squirrel Cuckoo) and widespread *cayana* group (e. Mexico to S America) are best treated as separate species; the latter group may include further cryptic species in S America.

AZTEC [OLIVE-THROATED] PARAKEET *Eupsittula [nana] astec* (p. 140). Differences in morphology, plumage, bare parts, and voice indicate that the *astec* group (Mexico to w. Panama) and *nana* group (Jamaica; Jamaican Parakeet) are best treated as separate species; also see Latta et al. (2012).

NORTHERN COLLARED TROGON *Trogon [collaris] puella* (p. 146). Differences in plumage and song indicate that the *puella* group (Mexico to Panama) and *collaris* group (e. Panama to S America) are best treated as separate species.

TROPICAL RINGED KINGFISHER *Megaceryle torquata* (p. 148). Differences in morphology, plumage, ecology, and voice indicate that the *torquata* group (Mexico to tropical S America) and *stellata* group (s. Chile and adjacent Argentina; Austral Ringed Kingfisher) are best treated as separate species.

GOLDEN-FRONTED WOODPECKER *Centurus aurifrons* (p. 154). Based on a limited mtDNA gene tree (García-T. et al. 2009), IOC split s. populations as Velasquez's Woodpecker *C. santacruzi* (se. Mexico to w. Nicaragua) from the *aurifrons* group (e. Mexico and s. Texas, US; Golden-fronted Woodpecker); yet another case of divergent mtDNA lineages being elevated to species with no biological support, and in this case with the conflicting real-world situation of an apparent broad hybrid swarm left unexplained.

GREEN JAY *Cyanocorax luxuosus* (p. 160). Differences in morphology, plumage, and voice suggest that the *luxuosus* group (e. Mexico and Texas, US; Texas Green Jay), *maya* group (s. Mexico to Honduras; Mayan Green Jay), and *vividus* group (w. Mexico; Western Green Jay) may best be treated as separate species.

WEDGE-TAILED SABREWING *Pampa curvipennis* (p. 166). The Wedge-tailed Sabrewing complex comprises 3 lineages sometimes treated as species, as by IOC; if split, Belize birds would best be called Yucatan Sabrewing *P. [c.] pampa*, but we find published arguments unconvincing.

STRIPE-TAILED HUMMINGBIRD *Eupherusa eximia* (p. 166). Differences in morphology, plumage, and voice suggest that the *eximia* group (Mexico to Nicaragua; Northern Stripe-tailed Hummingbird) and *egregia* group (Costa Rica to Panama; Southern Stripe-tailed Hummingbird) may represent separate species.

CANIVET'S EMERALD *Cynanthus canivetii* (p. 168). As explained by Howell (1993a), Salvin's Emerald *C. [c.] salvini* is distinct from Canivet's Emerald, and both were treated as species by Howell & Webb (1995). The ranges of the 2 taxa approach (or even overlap?) in Guatemala and adjacent Mexico, but with no evidence of hybrids.

WHITE-CHINNED SWIFT *Cypseloides cryptus* (p. 172). Critical study needed of birds in n. Central America, and of the enigmatic taxon *storeri* in Mexico, to resolve species limits in the White-chinned Swift complex (cf. Howell 1993b).

RICHMOND'S SWIFT *Chaetura [vauxi] richmondi* (p. 174). Differences in morphology, plumage, voice, and ecology indicate that the resident *richmondi* group (Middle America) and migratory *vauxi* group (N America; Vaux's Swift) are best treated as separate species; status of smaller birds in n. Yucatan Peninsula (*gaumeri*) is unclear.

CAVE SWALLOW *Petrochelidon fulva* (p. 176). Differences in morphology, plumage, and voice suggest that the *pallida* group (n. Mexico and sw. US; Northern Cave Swallow) and *citata* group (se. Mexico; Yucatan Cave Swallow) may be separate species; status of Caribbean taxa in Cave Swallow complex also in need of study.

RIDGWAY'S ROUGH-WINGED SWALLOW *Stelgidopteryx [serripennis] ridgwayi* (p. 176) Differences in morphology, plumage, ecology, and voice between the *ridgwayi* group (Mexico to Belize) and *serripennis* group (N America to Costa Rica; Northern Rough-winged Swallow), plus local sympatry and no evidence of interbreeding, indicate separate species are involved.

NORTHERN WEDGE-BILLED WOODCREEPER *Glyphorynchus [spirurus] pectoralis* (p. 182). Differences in plumage and voice indicate that the *pectoralis* group (Mexico to nw. S America) is best treated as a species distinct from other S American populations of the Wedge-billed Woodcreeper complex, which may comprise further cryptic species.

GRAYISH [OLIVACEOUS] WOODCREEPER *Sittasomus [griseicapillus] griseus* (p. 182). Differences in plumage and voice indicate that the *griseus* group (Mexico to n. S America) is best treated as a species distinct from other S American populations of the Olivaceous Woodcreeper complex, which comprises further cryptic species.

NORTHERN SPOTTED WOODCREEPER *Xiphorhynchus erythropygius* (p. 180). Differences in plumage and voice indicate that the *erythropygius* group (Mexico to Nicaragua) and *aequatorialis* group (Costa Rica to n. Ecuador; Southern Spotted Woodcreeper) are best treated as separate species.

PLAIN XENOPS *Xenops [minutus] genibarbis* (p. 182). Differences in plumage and voice indicate that the widespread *genibarbis* group (Plain Xenops, which may contain further cryptic species) is best treated as a species distinct from the *minutus* group of se. Brazil (White-throated Xenops).

MIDDLE AMERICAN [BUFF-THROATED] FOLIAGE-GLEANER *Automolus [ochrolaemus] cervinigularis* (p. 184). Differences in plumage and voice indicate that Buff-throated Foliage-gleaner comprises at least 2 species: the *cervinigularis* group (Mexico to nw. Panama) and *ochrolaemus* group (cen. Panama to S America), in addition to the recently split Chiriqui Foliage-gleaner *A. [o.] exsertus* (Costa Rica to Panama).

MIDDLE AMERICAN [BRIGHT-RUMPED] ATTILA *Attila [spadiceus] flammulatus* (p. 210). Differences in plumage and voice suggest that the *flammulatus* group (e. Mexico to w. Ecuador) may best be treated as species distinct from the *pacificus* group (w. Mexico; West Mexican Attila), and *spadiceus* group (S America, e. of Andes) .

NORTHERN OCHRE-BELLIED FLYCATCHER *Mionectes [oleagineus] assimilis* (p. 196). Differences in voice, plumage, and morphology suggest that the *assimilis* group (Mexico to n. S America), *pacificus* group (sw. Colombia to nw. Peru), and *oleagineus* group (S America, e. of Andes) may best be treated as separate species.

MISTLETOE TYRANNULET *Zimmerius parvus* (p. 198). Distinct vocal differences between formally undescribed northern birds (Belize to Costa Rica; Northern Mistletoe Tyrannulet), 'central' birds *Z. [p.] parvus* (Costa Rica to cen. Panama; Southern Mistletoe Tyrannulet), and undescribed southern birds (e. Panama; Darien Mistletoe Tyrannulet) suggest that three cryptic species may be involved (Howell & Dyer 2022).

MESOAMERICAN [GREENISH] ELAENIA *Myiopagis [viridicata] placens* (p. 198). Differences in voice and morphology indicate that the Greenish Elaenia complex comprises multiple cryptic species, also including *M. [v.] minima* (endemic to w. Mexico; West Mexican Elaenia) and *M. [v.] accola* (Costa Rica to nw. S America; Colombian Elaenia), plus others in S America.

SCLATER'S [YELLOW-OLIVE] FLATBILL *Tolmomyias [sulphurescens] cinereiceps* (p. 198). Differences in voice and morphology indicate that the Yellow-olive Flatbill complex comprises multiple cryptic species, among which the *cinereiceps* group (Mexico to Costa Rica) is distinct from populations ranging n. to Panama.

NORTHERN TROPICAL PEWEE *Contopus [cinereus] bogotensis* (p. 204). Differences in voice and plumage indicate that the *bogotensis* group (Mexico to n. S America) and *cinereus* group (e. and s. S America; Southern Tropical Pewee) are best treated as separate species, perhaps with further cryptic species involved.

BLACK PHOEBE *Sayornis nigricans* (p. 206). Differences in voice and plumage suggest that the *nigricans* group (N America s. to Panama; Northern Black Phoebe) and *latirostris* group (S America; Southern Black Phoebe) may represent separate species.

NORTHERN SOCIAL FLYCATCHER *Myiozetetes [similis] texensis* (p. 212). Differences in voice and morphology indicate that the Social Flycatcher complex comprises at least 5 cryptic species, also including *M. [s.] columbianus* (Costa Rica to nw. S America; Colombian Social Flycatcher); *M. [s.] grandis* (w. Ecuador to nw. Peru; Western Social Flycatcher); *M. [s.] similis* (Amazonian region; Amazonian Social Flycatcher); and *M. [s.] pallidiventris* (e. Brazil to n. Argentina; Austral Social Flycatcher).

NORTHERN STREAKED FLYCATCHER *Myiodynastes maculatus* (p. 210). Differences in voice and morphology indicate that the *maculatus* group (Mexico to S America) and *solitarius* group (breeds s. S America; Austral Streaked Flycatcher) are best treated as separate species.

MIDDLE AMERICAN [TROPICAL] KINGBIRD *Tyrannus [melancholicus] satrapa* (p. 212). Differences in voice (especially dawn song), morphology, and plumage indicate that the *occidentalis* group (w. Mexico; West Mexican Kingbird), *satrapa* group (e. Mexico to nw. S America), and *melancholicus* group (widespread in S America; South American Kingbird) may best be treated as separate species.

FORK-TAILED FLYCATCHER *Tyrannus savana* (p. 214). Differences in morphology, ecology, and provisionally in voice suggest that the *monachus* group (Mexico to nw. S America; Northern Fork-tailed Flycatcher) and *savana* group (breeds s. S America; Austral Fork-tailed Flycatcher) may represent separate species.

MAYAN [TROPICAL] MOCKINGBIRD *Mimus [gilvus] gracilis* (p. 216). Differences in voice and plumage suggest that the Tropical Mockingbird complex probably comprises multiple species, including the *gracilis* group of Mexico to Honduras (Howell 2019), which one molecular study suggested may be more closely related to Northern Mockingbird *M. polyglottos* than to other populations of Tropical Mockingbird (Lovette et al. 2012).

WHITE-THROATED THRUSH *Turdus assimilis* (p. 216). Vocal differences (most evident in call notes) between populations e. and w. of Isthmus of Tehuantepec, Mexico, hint that cryptic species may be involved; study needed.

MESOAMERICAN [BAND-BACKED] WREN *Campylorhynchus zonatus* (p. 220). Vocal and morphological differences indicate that the *zonatus* group (Mexico to n. Nicaragua) is best treated as specifically distinct from the *costaricensis* group (s. Nicaragua to w. Panama; Costa Rican Wren) and *brevirostris* group of nw. S America; also supported by molecular data (Vázquez-M. & Barker 2021).

HOUSE WREN *Troglodytes aedon* (p. 220). Differences in voice, plumage, and ecology suggest that the *aedon* group (N America; Northern House Wren), *brunneicollis* group (Mexico; Brown-throated Wren) and *musculus* group (s. Mexico to S America; Southern House Wren) may represent separate species, along with some Caribbean taxa in the House Wren complex.

GRASS [SEDGE] WREN *Cistothorus [platensis] elegans* (p. 220). We follow the study by Robbins & Nyári (2014) who advocated splitting the Sedge Wren complex into 8 species, including *elegans* of Middle America.

CAROLINA WREN *Thryothorus ludovicianus* (p. 222). Differences in voice and plumage suggest that the *albinucha group* (Mexico to n. Cen America; White-browed Wren) and *ludovicanus* group (N America to ne. Mexico; Carolina Wren) may represent separate species, as treated by some authors.

WHITE-BELLIED WREN *Uropsila leucogastra* (p. 220). Song differences suggest that the *pacifica group* (w. Mexico; Western White-bellied Wren) and *leucogastra* group (e. Mexico to Honduras; Eastern White-bellied Wren) may represent separate species.

PLAIN WREN *Cantorchilus modestus* (p. 220). Although usually now treated as 3 species (Cabanis's Wren, Canebrake Wren, Isthmian Wren) we find the case for splitting Plain Wren into 3 species unconvincing, another instance of divergent mitochondrial lineages elevated to the level of species by sophistry rather than science. With respect to possible vocal differences, the paper by Saucier et al. (2015) and Saucier's follow-up proposal to NACC were misleading at best, cf. Boesman (2016) and our own analysis. Song differences among taxa are not at all clear, although there may be some differences in scolding calls. Moreover, and perhaps unsurprisingly, Freeman & Montgomery (2017) found *zero* discrimination between 'Isthmian Wren' and 'Cabanis's Wren' in song playback experiments, yet their finding of much greater discrimination led to lumping of Scarlet-rumped Tanager taxa.

MIDDLE AMERICAN [WHITE-BREASTED] WOOD WREN *Henicorhina [leucosticta] prostheleuca* (p. 224). Vocal and morphological differences indicate that the *prostheleuca* group (Mexico to w. Colombia) is best treated as specifically distinct from other S American populations.

NORTHERN [LONG-BILLED/TRILLING] GNATWREN *Ramphocaenus [melanurus] rufiventris* (p. 224). Differences in voice and plumage (supported by molecular analysis; Smith et al. 2018) indicate that the *rufiventris* group (Mexico to nw. Peru) is best treated as specifically distinct from other S American populations of the *R. melanurus* complex

BLUE-GRAY GNATCATCHER *Polioptila caerulea* (p. 224). Differences in morphology, voice, and plumage suggest that the Blue-gray Gnatcatcher complex may comprise up to 5 cryptic species (as highlighted by Smith et al. 2018), 2 of them in Belize: resident Mexican Gnatcatcher *P. [c.] deppei* and migrant Eastern Gnatcatcher *P. [c.] caerulea*.

MAYAN [MANGROVE] VIREO *Vireo [pallens] semiflavus* (p. 228). Differences in voice, morphology, and ecology indicate that the *semiflavus* group (Mexico to n. Guatemala) and *pallens* group (Mexico to Costa Rica; Mangrove Vireo) are best treated as separate species; differences between them are as great or greater than between other vireo taxa treated as species, such as the Red-eyed/Chivi Vireos *V. olivaceus/V. chivi* and Cassin's/Plumbeous Vireos *V. cassini/V. plumbeus*.

GRAY-HEADED VIREO *Vireo [plumbeus] notius* (p. 228). Differences in voice, morphology, and plumage indicate that the *notius* group (s. Mexico to Honduras) and *plumbeus* group (interior w. N America s. to Mexico; Plumbeous Vireo) are best treated as separate species. The study splitting N American 'Solitary Vireo' taxa into 3 species was limited to US populations and did not consider these distinctive birds, which are sometimes known as Central American Vireo. That name, however, misleadingly implies a much larger geographic range; Gray-headed Vireo juxtaposes well with Blue-headed Vireo, the only other member of the Solitary Vireo complex with which Gray-headed co-occurs.

STRIPE-CROWNED [GOLDEN-CROWNED] WARBLER *Basileuterus culicivorus* (p. 234). Differences in morphology and voice indicate that the *culicivorus* group (e. Mexico to Panama) is best treated as a species separate from other taxa in the Golden-crowned Warbler complex in S America, and perhaps also from birds in w. Mexico.

OLIVE SPARROW *Arremonops rufivirgatus* (p. 248). Differences in morphology, plumage, and song suggest that Olive Sparrow may comprise multiple species, including the *verticalis* group (se. Mexico to Belize; Yucatan Olive Sparrow), and 3 other groups in Mexico plus the *superciliosus* group (Honduras to Costa Rica; Southern Olive Sparrow); see under Green-backed Sparrow, following.

GREEN-BACKED SPARROW *Arremonops chloronotus* (p. 248). Based on plumage, song, and ecology, the enigmatic taxon *twomeyi* of Honduras appears to belong to the Southern Olive Sparrow group, despite inexplicably having been described (and still treated) as a race of the rather different Green-backed Sparrow *A. chloronotus*.

ORANGE-BILLED SPARROW *Arremon aurantiirostris* (p. 248). Differences in call and song, coupled with minor plumage differences, suggest that the *rufidorsalis* group (Mexico to nw. Panama; Northern Orange-billed Sparrow), *aurantiirostris* group (Costa Rica to nw. Peru; Western Orange-billed Sparrow), and *spectabilis* group (s. Colombia to n. Peru, e. of Andes; Eastern Orange-billed Sparrow) may represent separate species.

MIDDLE AMERICAN BUSH-TANAGER *Chlorospingus [flavopectus] ophthalmicus* (p. 248). Dawn songs of the *ophthalmicus* group (Middle America) are distinctly high-pitched vs. the low-pitched songs of the *flavopectus* group (S America), supporting species status for these 2 groups; within each region there are likely further cryptic species, but biological data (notably recordings of dawn songs) are sparse or lacking for many regions and taxa.

RED CROSSBILL *Loxia curvirostra* (p. 254). Taxonomy worldwide of the Red Crossbill complex remains vexed, and much study is needed. Numerous different call types have been identified and Belize birds appear to be Type 11, known as Middle American [Red] Crossbill.

NORTHERN BLACK-FACED GROSBEAK *Caryothraustes poliogaster* (p. 258). Differences in voice and plumage indicate that the *poliogaster* group (Mexico to w. Honduras) and *scapularis* group (e. Honduras to Panama; Southern Black-faced Grosbeak) are best treated as separate species. Mexican birds ignore Costa Rica recordings but respond strongly to their own vocalizations (R. C. Hoyer, pers. comm.), and vocal differences are appreciably greater than, say, those between Rose-breasted and Black-headed Grosbeaks in N America, which are treated as species.

EASTERN BLUE BUNTING *Cyanocompsa parellina* (p. 260). Differences in voice, plumage (and perhaps molt strategy?) indicate that the *indigotica* group (w. Mexico; Western Blue Bunting) and *parellina* group (e. Mexico to Nicaragua) are best treated as separate species.

BLUE SEEDEATER *Amaurospiza concolor* (p. 260). Based purely on genetic lineages, IOC split the *concolor* group (s. Mexico to Panama) and *aequatorialis* group (sw. Colombia to n. Peru) as species. Morphology and voice of all *Amaurospiza* are similar, and Ecuadorian birds respond aggressively to songs from Mexican birds (Howell, pers. obs.). Moreover, samples used in studies to date are very small, and it seems equally plausible to subsume most or all taxa as subspecies of a widespread Blue Seedeater (Howell & Dyer 2022).

MIDDLE AMERICAN [RED-CROWNED] ANT-TANAGER *Habia [rubica] rubicoides* (p. 262). Differences in voice and morphology indicate that the *rubicoides* group (e. Mexico to Panama) is best treated as a species separate from populations of the Red-crowned Ant-tanager complex in S America.

RED-THROATED ANT-TANAGER *Habia fuscicauda* (p. 262). Seemingly minor differences in voice and morphology suggest that the *salvini* group (Mexico to n. Nicaragua; Salvin's Ant-tanager) and *fuscicauda* group (s. Nicaragua to n. Colombia; Cabanis's Ant-tanager) may represent separate species.

BLACK [VARIABLE] SEEDEATER *Sporophila corvina* (p. 270). Differences in plumage and voice indicate that the *corvina* group (Mexico to Panama) and *ophthalmica* group (Costa Rica to nw. Peru; Hick's Seedeater) are best treated as separate species, with a narrow, apparently stable hybrid zone in cen. Panama (Olson 1981).

COMMON BANANAQUIT *Coereba [flaveola] luteola* and **COZUMEL BANANAQUIT** *Coereba [flaveola] caboti* (p. 272). Differences in voice and morphology suggest that the *luteola* group (mainland Americas) and *caboti* group (Cozumel I. to n. Belize) are separate species, both distinct from the *flaveola* group (Caribbean region), which likely comprises further multiple species.

BLUE-GRAY TANAGER *Tangara episcopus* (p. 274). Striking plumage differences between the *cana* group (Mexico to nw. Peru; Northern Blue-gray Tanager) and *episcopus* group (S America, e. of Andes; Amazonian Blue-gray Tanager) suggest separate species may be involved.

GRAY-HEADED TANAGER *Eucometis penicillata* (p. 276). Differences in morphology and voice between the *spodocephalus* group (Mexico to nw. S America; Northern Gray-headed Tanager) and *penicillata* group (widespread in S America) suggest they may represent separate species.

YELLOW-BILLED CACIQUE *Amblycercus holosericeus* (p. 276). Differences in voice and morphology suggest that the *holosericeus* group (Mexico to w. Peru; Western Yellow-billed Cacique) and *australis* group (Colombia to Bolivia, e. of Andes; Eastern Yellow-billed Cacique) may represent separate species.

REFERENCES

Publications with 6 or more coauthors are cumbersome to cite and are simply noted as, for example, 'Yew, F., & 5 coauthors.'

American Ornithologists' Union (AOU). 1998. Check-list of North American Birds, 7th edition. AOU, Washington, DC.

Boesman, P. 2016. Notes on the vocalizations of Plain Wren (*Thryothorus modestus*) and Canebrake Wren (*Thryothorus zeledoni*). HBW Alive Ornithological Note 294. In: Handbook of the Birds of the World Alive. Lynx Edicions, Barcelona.

Freeman, B. G., & G. A. Montgomery. 2017. Using song playback experiments to measure species recognition between geographically isolated populations: a comparison with acoustic trait analyses. Auk 134:857–870.

García-T., E. A., A. Espinosa de los Monteros, M. del Coro Arizmendi, & A. G. Navarro-S. 2009. Molecular systematics of the Red-bellied and Golden-fronted Woodpeckers. Condor 111(3):442–452.

Gill, F., D. Donsker, & P. Rasmussen (eds.). 2021. IOC World Bird List (v11.2). https://doi.org/10.14344/IOC.ML.11.2.

Howell, S.N.G. 1993a. Taxonomy and distribution of the hummingbird genus *Chlorostilbon* in Mexico and northern Central America. Euphonia 2:25–37.

Howell, S.N.G. 1993b. More comments on White-fronted Swift. Euphonia 2:100–101.

Howell, S.N.G. 2019. To split a mockingbird. https://ebird.org/news/to-split-a-mockingbird

Howell, S.N.G. 2021. What isn't a species? North American Birds 72:16–25.

Howell, S.N.G., & D. Dyer. 2022. Costa Rica—even richer than we thought? North American Birds 73:22–33.

Howell, S. N. G., & F. Schmitt. 2018. Birds of Chile: A Photo Guide. Princeton Univ. Press.

Howell, S.N.G., & S. Webb. 1995. A Guide to the Birds of Mexico and Northern Central America. Oxford Univ. Press.

Howell, S.N.G., & K. Zufelt. 2019. Oceanic Birds of the World: A Photo Guide. Princeton Univ. Press.

Howell, S.N.G., A. Jaramillo, N. Redman, & R. S. Ridgely. 2012. What's the point of field guides: taxonomy or utility? Neotropical Birding 11:16–21.

Jones, H. L. 2003. Birds of Belize. Univ. Texas Press.

Latta, S. C., A. K. Townsend, & I. J. Lovette. 2012. The origins of the recently discovered Hispaniolan Olive-throated Parakeet: a phylogeographic perspective on a conservation conundrum. Caribbean Journal of Science 46(2–3):143–149.

Lerner, H. L., M. C. Klaver, & D. P. Mindell. 2008. Molecular phylogenetics of the buteonine birds of prey (Accipitridae). Auk 304(2):304–315.

Lovette, I. J., & 10 coauthors. 2012. Phylogenetic relationships of the mockingbirds and thrashers (Aves: Mimidae). Molecular Phylogenetics and Evolution 63:219–229.

Monti, F., & 7 coauthors. 2015. Being cosmopolitan: evolutionary history and phylogeography of a specialized raptor, the Osprey *Pandion haliaetus*. BMC Evolutionary Biology 15:255. https://doi.org/10.1186/s12862-015-0535-6.

Olson, S. L. 1981. The nature of the variability of the Variable Seedeater in Panama (*Sporophila americana*: Emberizinae). Proc. Biol. Soc. Wash. 94:380–390.

Olson, S. L., & R. C. Banks. 2007. Lectotypification of *Larus smithsonianus* Coues, 1862 (Aves: Laridae). Proc. Biol. Soc. Wash. 120(4):382–386.

Oswald, J. A., & 8 coauthors. 2016. Willet be one species or two? A genomic view of the evolutionary history of *Tringa semipalmata*. Auk 133(4):593–614.

Robbins, M. B., & A. S. Nyári. 2014. Canada to Tierra del Fuego: species limits and historical biogeography of the Sedge Wren (*Cistothorus platensis*). Wilson Journal of Ornithology 126:649–662.

Russell, S. M. 1964. A distributional study of the birds of British Honduras. Ornithological Monographs no. 1, AOU.

Sánchez, C., & 12 coauthors. 2012. New and noteworthy records from northwestern Peru. Boletín Informativo UNOP 7(2):18–36.

Sangster, G., & 6 coauthors. 2011. Taxonomic recommendations for British birds: seventh report. Ibis 153:883–892.

Saucier, J. R., C. Sánchez, & M. D. Carling. 2015. Patterns of genetic and morphological divergence reveal a species complex in the Plain Wren (*Cantorchilus modestus*). Auk 132:795–807.

Smith, B. T., & 5 coauthors. 2018. Species delimitation and biogeography of the gnatcatchers and gnatwrens (Aves: Polioptilidae). Molecular Phylogenetics and Evolution 126:45–57.

Vallely, A. C., & D. Dyer. 2018. Birds of Central America. Princeton Univ. Press.

Vázquez-M., H., & F. K. Barker 2021. Autosomal, sex-linked, and mitochondrial loci resolve evolutionary relationships among wrens in the genus *Campylorhynchus*. Molecular Phylogenetics and Evolution 163:107242.

Wink, M., H. Sauer-Gürth, & H.-H. Witt. 2004. Phylogenetic differentiation of the Osprey *Pandion haliaetus* inferred from nucleotide sequences of the mitochondrial cytochrome b gene. Pp. 511–516 in R. D. Chancellor & B.-U. Meyburg (eds.). Raptors Worldwide. World Working Group on Birds of Prey, Berlin.

Yoon, C. K. 2009. Naming Nature: The Clash between Instinct and Science. Norton, New York.

INDEX OF ENGLISH NAMES